MW01201117

GT
SPORT

GOODYEAR
POLYGLAS - E70-14

MUSCLECAR
COLOR • HISTORY

Dodge Dart & Plymouth Duster

Steve Statham

MBI Publishing Company

First published in 2000 by MBI Publishing Company, 729 Prospect Avenue, PO Box 1, Osceola, WI 54020-0001 USA

MBI Publishing Company books are also available at discounts in bulk quantity for industrial or sales-promotional use. For details write to Special Sales Manager at Motorbooks International Wholesalers & Distributors, 729 Prospect Avenue, PO Box 1, Osceola, WI 54020-0001 USA.

Library of Congress Cataloging-in-Publication Data available
ISBN 0-7603-0760-1

On the front cover: With a 275-horsepower 340 V-8 under the hood, the 1970 Dodge Dart Swinger posted some impressive numbers. Designed strictly as a performance engine, the 340 was never even offered with a two-barrel carburetor. The Sublime model pictured is owned by Mike McFatridge of LaPorte, Texas.

On the frontispiece: The top-of-the-line 1969 Dart GTS featured the famed bumble bee stripe. The high-performance Scat Pack car was available in a hard top or a convertible.

On the title page: Like the Dart Sport, the 1973 Dusters were given new, chunkier nose and tail treatments. The deck spoiler is an owner addition from a 1972 model. The beloved 340 V-8 made its final appearance in the 1973 models, and though output dropped to 240 horsepower, it was still an admirable performer.

On the back cover: Top: The first generation Dart was given a mild facelift for 1966, with scalloped front fenders and a new grille being the major changes. Front bumper guards were a worthwhile option, as were disc brakes, which required the use of the larger 14-inch wheels to clear the rotors. ***Bottom:*** With its hood bulges and bumble bee stripes, one's eyes can not help but be drawn to the 1969 Dart GTS. With a 340 V-8 under the hood, the GTS took on the small-block competition and most often conquered it.

Edited by Paul Johnson

Designed by Arthur Durkee

Printed in China

Contents

Acknowledgments

During the course of preparing this book I ran into many A-body Mopar fans, several of whom I need to thank here. First, I'd like to thank Fred Schimmel, who labored in the Plymouth styling studios for 20 years, for sitting with me for interviews and for allowing me to photograph several of his design proposals. Those drawings were never turned into sheet metal, but at least people can enjoy them here, Fred.

I must likewise thank Norm Kraus, whose Mr. Norm's Grand Spaulding Dodge in Chicago produced some of the fastest, most memorable A-body cars to ever hit the streets. Norm graciously set aside time to share his memories with me, and I hope the reader finds his recollections as entertaining as I did. My dealings with everyone connected to Mr. Norm's Sport Club, such as President Jim Bodanis and Larry Weiner, were immensely satisfying and worthwhile. They may very well be the most professional automotive enthusiast's club from sea to shining sea.

I also owe a debt of gratitude to fellow automotive writer Rob Reaser, who graciously cracked open his photo files for film of Sam Hall's 1969 F-5 green Dart GTS 340 and Charles Provance's 1968 Super Stock Hemi Dart. Thanks also need to go out to noted Mopar expert Keith Rohm, whose 1968 GTS is pictured herein, for his help understanding the various technical aspects of the 383 Darts.

Gathering those old period photos people love to see is no easy task, so I've got to thank Bob and Dottie Plumer at Drag Racing Memories (200 N. Kalmia Ave., Highland Springs, VA 23075, 804-328-0680) for granting permission to run their vintage drag racing shots. A tip o' the hat is likewise due Art Ponder at the DaimlerChrysler Historical Collection for his efforts in getting me the archival A-body material.

Thanks also go out to Ron Chinn, with the Mopar Muscle cars of Austin club, for bird-dogging the local hot Mopars, and to Kenneth Bashrum for the same assistance finding Houston-area cars. Thanks also are due Greg Rager, editor of *High Performance Mopar*, and Jerry Pitt, editor of *Mopar Muscle*, specifically for their help while I was on my photo safari at the 1999 Mopar Nationals in Ohio, and also for good advice and for sharing their voluminous knowledge of Mopars.

I must also acknowledge previous books on the subject that were helpful in researching my own text, starting with: *Standard Catalog of Chrysler 1924–1990*, by John Lee; *The Dodge Story*, by Thomas A. McPherson; *The Complete Book of Stock-Bodied Drag Racing*, by Lyle Kenyon Engel; and *Iacocca, an Autobiography*, by Lee Iacocca. Also gotta love those tattered old yellow copies of the *N.A.D.A. Official Used Car Guide* I kept all these years. Thanks for bringing those home from the office back in the 1970s, Dad! With the rise of the Internet there are a lot more sources than just paper books, and I found www.mopars.com/DusterDemonDartSport/ to be a particularly helpful site for the Mopar enthusiast.

Finally, I must give extravagant thanks to the car owners who graciously allowed me to photograph their cars for this book. These fine people include: Steven Dykes, Austin, Texas, 1966 Dart GT; Alan Schroer, New Knoxville, Ohio, 1967 Dart GTS 383 convertible; Keith and Joy Rohm, Wapakoneta, Ohio, 1968 Dart GTS 383; Bill Vargoshe, Madison, Connecticut, 1969 Dart GTS 383 convertible; Jim Poncik, Bayou Vista, Texas, blue 1969 Dart GTS hardtop; Tim Dusek, Zionsville, Indiana, Moulin Rouge 1970 Duster 340; Mike McFatridge, Le Port, Texas, Sublime 1970 Dart Swinger 340 clone; Alan Spry, Abilene, Texas, 1970 AAR 'Cuda; Dennis Barnes, Elmhurst, Illinois, 1971 Demon 340 Six-Pack GSS; Brian Uccello, The Woodlands, Texas, Glacial Blue 1971 Duster 340; James A. St. Clair, Struthers, Ohio, red 1972 Dodge Demon; Marvin Wente, New Bremen, Ohio, red 1973 Plymouth Duster 340, yellow 1974 Plymouth Duster 360; Don and Kim Wilkinson, San Angelo, Texas, black 1973 Dart Sport 340. Additionally, there are a couple shots herein taken at the 1999 Mopar Nationals, where I didn't get a chance to meet the competitors, as they were busy smoking tires and banging gears. I hope we get to meet at some future car show. Thanks, all.

Preface

I can still remember when I bought my 1968 Dart GT. I was actually shopping for a Barracuda or Challenger, but the newspaper ad for the Dart was intriguing, so my dad and I went to investigate. The car had relatively low miles, was being sold by the original owner, and was straight and original. As Darts go, it was hardly a top-line model. It was a GT hardtop with a 318 V-8, automatic transmission, and fender-mounted turn signal indicators, but no air conditioning or power steering.

But I wanted it. I'd always loved those shiny, heavy, Mopar consoles running between the twin bucket seats. I liked the clean lines of the car, the GT trim, and the fact it still had faint traces of that unique, new-Mopar smell. And after the series of squishy Fords I'd recently owned, the Dart's GT suspension made the car feel as if it were ready to qualify at Daytona.

The car served me well for years. It carried me and my brother halfway across the country to the family farm in Georgia and back. Once, my friend Jim and I snuck his Datsun 280Z and my Dart GT into a recently closed San Antonio circle track for a few clandestine laps. We slid around the sunken, paved bowl, out of sight from the main road, laughing the whole time. Only a spark plug wire coming in contact with the exhaust manifold ended our fun that day.

What I learned in those years of Dart ownership is what a lot of Dart and Duster enthusiasts have known for a long time. You don't have to have a fire-breathing 426 Hemi underhood to have fun with your Mopar. Chrysler packed a lot of performance, economy, utility, and style into its A-body compact cars, especially during the musclecar era, when the Dart GTS, Demon 340, and Duster 340 were the equal of just about any other American performance car you could name.

The author's 1968 Dart GT, circa 1980.

Sure, most people remember the humble Dart as grandma's preferred mode of transport to the Sunday social. But had grandma looked a bit further down the options list, she might have discovered possibilities that were antisocial, yet fun, like the 383-ci V-8.

Introduction

Grandma's Hot Rod

All right, there's no getting around it. No sense pretending otherwise, and no value in wishful thinking. For fans of Mother Mopar's Dodge Dart and Plymouth Valiant, before we utter the first word about high-performance models or big-block V-8s, we may as well acknowledge the obvious. Out in the real world these sturdy cars were thought of, when they were thought of at all, as grandma's car. Thanks to conservative styling, economical engines, and an inexpensive price, the Dart and Valiant will forever be remembered by Americans as the sensible choice of the elderly fixed-income set.

But the rest of us can share our dirty little secret. The Dart, Demon, Valiant, and Duster may have been economy cars, but they were no draft dodgers when it came to musclecar duty in the 1960s. Some of the best factory musclecars of the era were based on the A-body platform. Additionally, their trim dimensions and light weight made them ideal raw material for the racetrack. Whatever the general public may think, the Dart and Valiant-based Duster had their moments in the sun as genuine high-performance cars, and to this day they have their own loyal following.

The point where the Valiant and Dart stopped being strictly sensible and completed their transformation into honest musclecars was 1967. Prior to that the Dart had been available with a small but powerful 273-ci V-8, but in 1967 Dodge shoe-horned in the big-block 383-ci V-8. There was no logical reason to cram that much engine into so small a car, but horsepower fever had beset Detroit's auto manufacturers, and no cars were immune. Ford's Mustang ponycar could be ordered with a 390-ci V-8, Chevrolet's trim Camaro with a 396, and Pontiac's Firebird with a 400. Chevy's Nova SS sported a high-winding 327 V-8 on its options list.

Suddenly, the Dart's GT package wasn't enough. Dodge added a GTS model to the line-up, and the big chrome letters on the fenders signified speed and power. The 340-ci small-block V-8 that debuted in the GTS in 1968 eventually became so closely identified with the Dart and Duster that the engine and car are usually spoken as one word, "340Duster." Chrysler Corporation eventually even built a handful of race-only Hemi-powered Darts and A-body Barracudas. In later years, Darts and Dusters sprouted hood scoops, spoilers, and the cartoonish colors and graphics that were a staple of 1970s-era performance cars. By the time the early 1970s rolled around, the Dart/Duster combo had joined the long line of American economy cars that had been made special by insertion of too much engine.

The Valiant and Dart earned loyal followers around the world, some of whom enjoyed their own performance versions of the car—notably in Australia. [See chapter 6 sidebar, Worldwide Fans.] Unique variants were also built for the Australian and South American markets. The A-body's exploits reach far beyond U.S. shores.

In today's musclecar market, the Darts and Dusters are often overshadowed by the larger, intermediate-sized Mopars. Fewer A-body performance cars get the full restoration treatment the more popular B-body (Charger, Super Bee, Road Runner, GTX) and E-body (Challenger, post-1970 Barracuda) Mopars receive. In part, this is due to economics. Restored Darts, Dusters, and Demons do not yet bring in the big money that other musclecars command. Yet their restoration costs are often a little lower than the big-inch monsters. Traditionally, the monster-engine big-block cars command the highest prices. High-winding small-block compacts rarely set the collector car market afire. (Of course, the side benefit of this is that Dart and Duster musclecars in decent shape have remained one of the few economical choices for the cash-poor musclecar enthusiast. Additionally, A-body cars that were considered compact in the 1960s feel downright large by today's standards.)

Now that DaimlerChrysler has given the entire Plymouth nameplate the ax, who knows how future generations will view old cars like the Duster? In the here and now, however, the A-body cars still have their fans. Groups like the Slant Six Club of America, the Dart GTS Registry, Mr. Norm's Sport Club, and scores of local Mopar clubs count Darts and Dusters among their membership. The Mopar Nationals and Chryslers at Carlisle car shows bulge with A-body entries. The cars are still regularly found at drag strips across America. A quick search on the Internet will reveal a considerable number of international fans. Even grandma would have to be impressed.

one
Stirrings of Performance:
GTs and V-8s, 1960–1966

The idea that Chrysler Corporation would soon unleash an army of high-performance compact cars into the world did not seem very likely in 1960. With the new decade came a shift in the American car market, brought about by assorted converging factors. Small imported cars in general, and the Volkswagen Beetle in particular, had made noticeable inroads into the U.S. market. Their economical engines, low prices, and relatively restrained styling had earned them a growing fan base. Additionally, the economic recession of 1958 had made Americans more aware of the virtues of low-cost transportation. American Motors was enjoying good times, thanks to strong sales of its bare-bones Rambler.

The Valiant was Chrysler's response to those factors, and few at the corporation were thinking of a high-performance future for the car (although they were eager to showcase the car's new engine). The first Valiants offered no V-8 option, no four-speed transmission, no axle-twisting rear end ratio. Economy was the focus. The Valiant's wheelbase was 11.5 inches shorter than that of the standard Plymouth, and overall length was 25 inches less. Shipping weight was down roughly 700 to 900 pounds compared to the larger models.

The Dart was given a mild facelift for 1966, with scalloped front fenders and a new grille being the major changes. Front bumper guards were a worthwhile option, as were disc brakes, which required the use of the larger 14-inch wheels to clear the rotors.

Although saddled with odd styling, the new 1960 Plymouth Valiant did bring some solid engineering to market. The Valiant had a unitized body, used a torsion bar front suspension, and was motivated by the cleverly designed Slant Six engine. In addition to its long-runner intake manifold, the Slant Six used an alternator, at a time when most other American engines relied on generators. *Reprinted with the permission of the DaimlerChrysler Corporate Historical Collection*

The first high-performance Valiants relied on the "Hyper-Pak" version of the 170-ci Slant Six. The Hyper-Pak bumped up the Slant's power by the use of a four-barrel carburetor on an elongated manifold, an increase in the compression ratio, free-flowing exhaust manifolds, and a hotter cam. This engine helped the Valiant to dominant wins in NASCAR's short-lived compact class and in NHRA and AHRA class competition. *Reprinted with the permission of the DaimlerChrysler Corporate Historical Collection*

The Valiant actually entered the market designated as its own brand, to be sold at any Chrysler Corporation dealer. Although a solid car, the Valiant was saddled with odd styling. The look, guided by Chrysler stylist Virgil Exner, featured dramatically accentuated fender lines, an open, airy greenhouse, and a distinctive spare tire indentation on the trunk lid. The Valiant was initially offered with two trim levels, the stripped V-100 and high-end V-200. Body styles were limited to a four-door sedan and six- or nine-passenger station wagons.

The Valiant's American competition for 1960 was also all new—the Ford Falcon and Chevrolet Corvair. (And, to a lesser extent, the Studebaker Lark, introduced in 1959.) Each was a unique interpretation of what an American economy car should be— compact, affordable, and powered by a new six-cylinder engine. The similarities, however, ended there.

Of the three, the Falcon followed the most conventional path. Although its trim, overhead-valve, pushrod, inline six was an improvement over Ford's previous flathead six-cylinders, it just met the bare minimum requirements to be considered a modern engine. The car's suspension was no technological breakthrough, and the Falcon's styling established no new trends.

The Corvair was the most technically intriguing of the three. With a rear-mounted, air-cooled, flat six for power, the Corvair was also the most import-like of Detroit's new economy offerings. The Corvair boasted an independent coil-spring suspension and distinctive styling as well.

The Valiant fit somewhere in the middle. Larger than its competitors, the Valiant definitely offered a distinctive appearance, although whether that was a plus or a minus remains an open question. But the Valiant's new engine, the inline "Slant Six," was well designed and proved to be a mainstay of the Chrysler line-up for nearly three decades.

The Slant Six was so named because the engine was tilted 30 degrees to one side. This allowed a low hoodline, and an additional bonus in the power department. The Slant Six's intake manifold featured six individual, gently curved runners. The long runners promoted a smooth air/fuel mixture, and provided a mild ram-tuning effect for good midrange power. Most sixes of the period used blocky, log-type intake manifolds—economical to produce, but full of sharp curves and restrictive passages that hampered the engine's breathing abilities.

The Slant Six debuted in the Valiant at 170 cubic inches, and used an alternator rather than a generator, a first for Chrysler. It had an 8.5:1 compression ratio and a single-barrel carburetor. Unlike most engines, its bore was larger than its stroke, which resulted in a quick-revving package. The 170-ci Slant Six was rated at 101 horsepower, compared to 80 or 95 for the Corvair's six and 90 for the Falcon's. Put up against other American Sixes of the time, it was practically a high-performance motor.

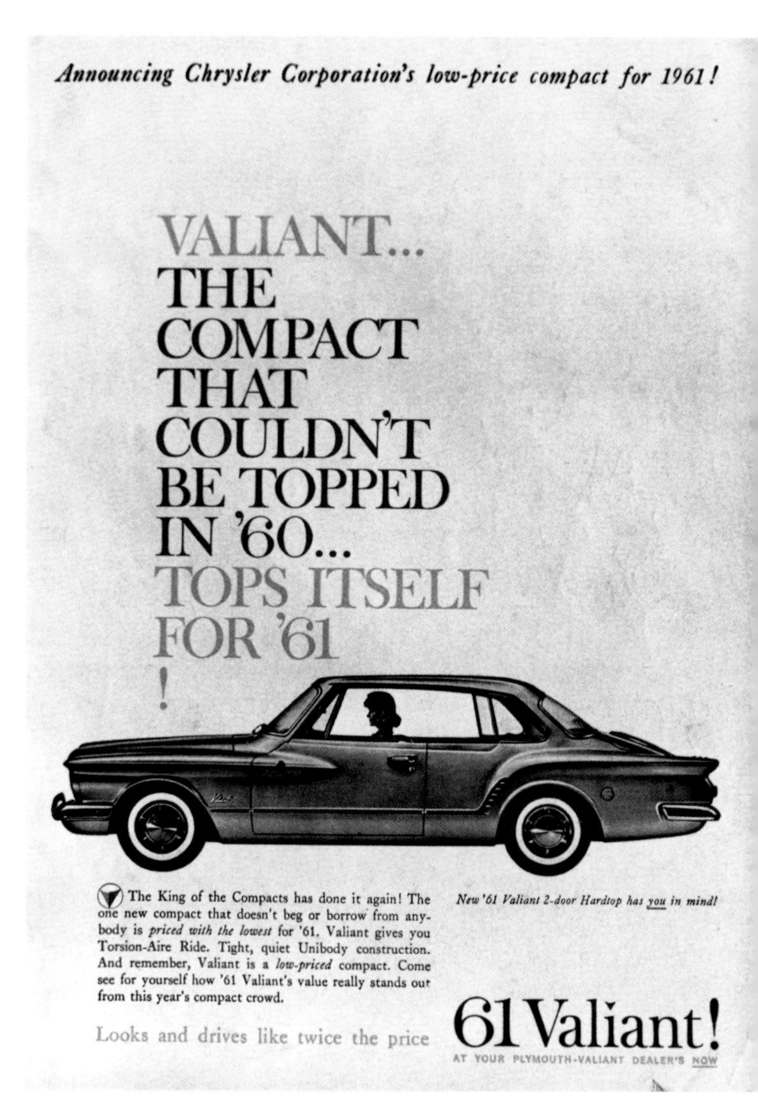

As performance aspirations go, the changes for 1961 were modest, but at least a new two-door hardtop model was added to the Valiant line. That year a slightly altered version of the Valiant was sold by Dodge dealers as the Lancer.

Adding to the powertrain's gee-whiz factor was the optional automatic transmission, which featured push-button gear selection. A row of buttons was lined up vertically on the left side of the instrument panel. The automatic was a special downsized and lightened unit designed specifically for the Valiant.

Popular Mechanics recorded a 0–60 mile per hour time of 14.5 seconds for the Valiant in its January 1960 issue, while the Corvair and Falcon took

Dodge considered keeping the Lancer name for its new 1963 compact, but instead chose the Dart name, which had been used previously on the larger, entry-level Dodges. Although a corporate stablemate of the Plymouth Valiant, the Dart rode on a 111-inch wheelbase, while the smaller Valiant wheelbase stretched only 106 inches. The top Dart for 1963 was the GT; the hottest powertrain combination was a 145-horsepower 225-ci Slant Six and three-speed manual transmission. *Reprinted with the permission of the DaimlerChrysler Corporate Historical Collection*

Dart GT styling was touched up in 1964, with a new convex-shaped grille replacing 1963's concave shape. A high belt molding and small hood scoop molding rounded out the visible changes. Out of sight was a much bigger change—an optional 273-ci two-barrel V-8. *Reprinted with the permission of the DaimlerChrysler Corporate Historical Collection*

Chrysler Corporation gave the high-performance 273 a spit-shined look to go with the newfound 235 horsepower. A chrome air cleaner and black crinkle-finish valve covers provided the polish. *Reprinted with the permission of the DaimlerChrysler Corporate Historical Collection*

Continued from page 13

16.5 seconds and 21.1 seconds, respectively, to accomplish the same feat. "There's none of the small-car feel that many want and many more don't," noted the *Popular Mechanics* crew of the Valiant. "It has a lot of power, plenty of speed and gives a secure feeling."

To show off the potential of the Slant Six, Chrysler engineers developed a Hyper-Pak parts package to bolster performance. (Internal Dodge and Plymouth technical material alternately refers to this engine as either "Hyper-Pak," "Hyper-Pac," or even "Hyper-Pack." Hyper-Pak is used here, as it seems to have caught on among enthusiasts.) The Hyper-Pak group included a four-barrel carburetor perched on a long-runner intake manifold and a 10.5:1 compression ratio. With the modifications, horsepower jumped to 148. The Hyper-Pak engines helped Valiants clean up at a NASCAR Daytona compact class stock car exhibition race in 1960.

The 1960 Valiant was outsold by both the Ford Falcon and Chevrolet Corvair, but still posted respectable production of 194,292 cars, a substantial percentage of overall Plymouth sales that year. The upscale V-200 four-door proved to be the most popular model.

In 1961 the Valiant was officially made a Plymouth model in the United States, although it remained a separate car line in overseas markets. A two-door hardtop also joined the line-up that year, giving buyers a somewhat sporty choice. Two-tone paint options encouraged further image building. A less-than-successful restyling of the full-size Plymouth allowed the Valiant four-door to move into the lead in Plymouth sales. That year the corporation also decided to spread the platform around, as Dodge got a version of the Valiant, named the Lancer. The Lancer had its own distinct grille and different trim, but was otherwise a Valiant clone. Dodge produced 74,774 Lancers in 1961.

For 1962 Chrysler's new economy cars started evolving a bit, although for the Lancer it proved to be an evolutionary dead end—the name was axed at the end of the model year. But a Lancer GT model debuted, pointing the direction for future compact Dodges. Despite the arrival of the GT, however, sales of the Lancer dropped nearly 50 percent from 1961.

The Valiant developed a sportier personality in 1962 as well. Plymouth's version of a sporty compact was the Valiant Signet-200. It, too, was sold as a two-door hardtop, with a unique grille treatment, Signet-200 identification, and bucket seats. The 225-ci Slant Six was optional.

These early 1960–62 compacts from Chrysler established that there was indeed a viable market for lively American economy sedans. Although certainly not performance cars in the traditional sense, the Valiant and Lancer proved that manufacturers could add a bit of flair to the mix and sell the results to the public.

Sporting Intentions

Of Detroit's new-for-1960 economy cars—Falcon, Corvair, and Valiant—Ford's pedestrian Falcon was the best seller. At Plymouth, securing third place was nothing unusual. At Chevrolet, however, where hopes for the Corvair ran high, being trounced by the Falcon was not how events were supposed to play out.

One way Chevrolet fought back was with a special Corvair named Monza. The Monza was no stripped economy car. It came standard with bucket seats and special trim. In 1961 Chevrolet offered the Monza with a four-speed transmission, at a time when all other American compacts plodded along with three-speeds. In 1961 the Monza became the best-selling Corvair model, a title it held until the car's retirement

The 1965 Valiant could be optioned into a fun car, but most of Plymouth's performance marketing was reserved for the Barracuda. The sportiest available Valiant for 1965 was the convertible with optional 273-ci V-8. *Reprinted with the permission of the DaimlerChrysler Corporate Historical Collection*

Funsize!

Dart '65
The Dodge-size compact. Big?
Sure, big.
That's Dart all over.
But compact. Load it.
Go ahead, really load it.
Now some people . . . lots of 'em.
Hard to use up all that space, isn't it?
'65 Dart
Roof line . . . rakish.
Spirit . . . mildly wild.
Under the hood? Zoorch.
Her? She likes fun.
She's with us.
What's holding you back?
Dodge comes on big
for '65. Join us.

'65 Dodge Dart

DODGE COMES ON BIG FOR '65 • DART • CORONET • POLARA • CUSTOM 880 • MONACO

Even in 1965 Dodge was promoting the Dart as a big, er, "Funsize!" compact. The 1965 Dart received a sheet metal freshening for the new year, with a more sharply squared-off hood, grille, and fenders.

The 1966 Dart GT with V-8 usually found itself competing in the C/Modified Production and D/Modified Production classes at the drag strip. Its natural competition was the Chevy II, Barracuda, and 289 Fairlane.

in 1969. Chevrolet even rolled out a turbocharged Monza Spyder from 1962 to 1964. Suddenly, sporty compacts seemed the wave of the future.

Ford's response was the Falcon Futura in 1961, followed by the Falcon Futura Sprint in 1963. The 1963 Sprint offered something the Corvair could not match—a V-8 engine option. Ford's new lightweight, 260-ci small-block V-8 proved perfect for the compact Falcon.

Chrysler Corporation also took a stab at the sporty compact car market. The 1960 Valiant offered the Hyper-Pak Slant Six option, which was by far the most powerful engine offered in an American compact. But the Valiant lacked the sporty accouterments of the Monza. The Hyper-Pak was fine at the racetrack, but Plymouth lacked a special model that could pull young people in off the street.

In 1962 Plymouth addressed that shortcoming with Valiant Signet and Dodge Lancer GT models. The Lancer GT was sold as a two-door hardtop only, and the car came standard with the larger 225-ci Slant Six.

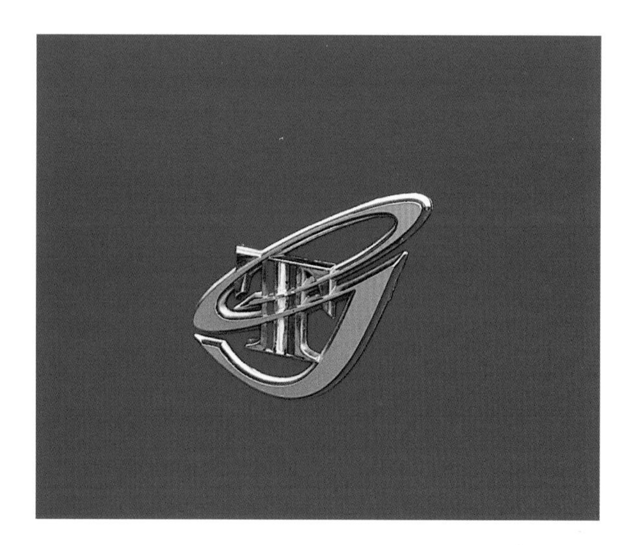

The 1965 Dart GT was easily identified by its heavily stylized GT script on the C-pillar, although sometimes people had to squint to make out the letters.

Barracuda—Swimming With the Tide

While Dodge decided to mold the Dart into a performance car, the Plymouth Division tried a different strategy. Using the Valiant as raw material, the company created a separate, distinct model, named the Barracuda. With dramatic styling and performance options aplenty, the Barracuda nailed down the performance corner of the market for Plymouth. The average Valiant, meanwhile, remained a frugal economy car.

Introduced April 1, 1964, as the Valiant Barracuda, these early cars were not much more than Valiants with huge glass fastbacks and a unique grille. The Barracuda did, however, beat Ford's Mustang to market by two weeks, and can legitimately lay claim to being the first entry in the "ponycar" class. First out of the gate didn't prove to be much advantage for the Barracuda, though, as the car was handily outsold by the Mustang.

As a Valiant offshoot, the Barracuda shared its powertrain options. The 170-ci Slant Six was standard, and the 225 six optional. The car could also be fitted with Chrysler's new 273-ci, 180-horsepower V-8 and a four-speed transmission. Bucket seats were standard equipment though, as was a folding rear seat.

The Barracuda stepped up a level in 1965 with the introduction of the Rallye Pack and Formula S options. The Rallye Pack featured a heavy-duty suspension and a Rallye stripe that advertised the car's performance intent. The Formula S came standard with a new high-performance version of the Commando 273-ci V-8 that produced 235 horsepower, thanks to a four-barrel carburetor, solid lifters, and free-flow exhaust. The package also included heavy-duty springs, shocks, and torsion bars, a sway bar, wider wheels, and Goodyear Blue Streak tires. The 1965 Formula S Barracuda was the first version of the fish that could be legitimately labeled a musclecar.

The 1966 Barracuda was mildly upgraded. The earlier car's small, skinny bumpers were replaced by larger bumpers with integrated front parking lamps. The grille was redesigned, as were the taillamps, and the front fenders were reshaped to good effect. On the inside, the instrument panel was redesigned and a new, full-length console replaced 1965's undersized shifter housing.

The Barracuda was given handsome new sheet metal for 1967. Stylist Milt Antonick's design, besides

To make the 1964 Barracuda, Plymouth stylists took the basic Valiant package and stretched and pulled it into a unique silhouette. The original Barracuda's fastback required the creation of the largest single piece of curved glass in an automotive application up to that point. The A-body Barracuda, a close sibling of the Valiant and Dodge Dart, was restyled for 1967 and given a more pleasing shape. *Reprinted with the permission of the DaimlerChrysler Corporate Historical Collection*

In 1970 the Barracuda name was transferred to a new E-body coupe that shared its heritage with the Dodge Challenger. The new cars were a rather late response to the ponycar wave the Ford Mustang had started, but were capable performers, as exemplified by the AAR 'Cuda shown here. The Duster assumed the role in the Plymouth A-body line-up previously held by the Barracuda.

being good looking, was also more practical to produce. The original Barracuda's huge, curved, one-piece rear glass was a nightmare to manufacture, and corporate decision-makers forbade anything like it for the next generation. "They had specified before we went into that, no way would we ever get a big back window like that first one," recalled Plymouth stylist Fred Schimmel. "They had horrible (production) problems on that." The new fastback was joined by a convertible model and a nicely proportioned "notchback" hardtop coupe, expanding the Barracuda line.

Performance features for the 1967 Barracuda included a 383 V-8 option, which the Dodge Dart GTS mimicked later in the year. In 1968 the new 340-ci small-block V-8 was a Barracuda and Dart GTS exclusive. Underrated at 275 horsepower, the lightweight, compact, high-winding 340 was a perfect fit for the Barracuda. The 383 Barracudas tended to be nose-heavy, and the more compact 340 engine offered equal or better performance while preserving ideal handling characteristics.

A series of 426-Hemi-powered 1968 Barracudas was made available to select racers. In 1969 Plymouth created even more identifiable models around the performance engines, the 'Cuda 340 and 'Cuda 383. The 'Cuda name was a natural, since that was the car's name on the street anyway. The 'Cuda came with two small, nonfunctional hood scoops and more noticeable striping.

Later in 1969, just as with the Dart, limited production runs of 440 models were created, assembled by Hurst Performance. The Valiant, meanwhile, dropped into the slow lane, devoid of the sort of performance options that were driving the Barracuda and Dart in exciting new directions.

The Barracuda was given a fresh design in 1970, based on the E-body platform, which was shared with the new Dodge Challenger. The E-body 'Cuda was one of the handsomest musclecars to issue from Detroit in the 1970s, and one of the most capable performers. The 1970 Barracuda line was available with virtually the entire run of Chrysler engines, from the 225-ci Slant Six to 318-ci and 340-ci small-block V-8s, up to the 383-ci, 440-ci, and 426-ci big-block V-8s. Only the 198-ci Six was not available that first year, although buyers had that choice, too, in 1971. The Hemi 'Cuda had the distinction of being the smallest, lightest street car to offer the 426 engine option, making it one of the most revered musclecars of the era.

The E-body 'Cuda did have some drawbacks compared to the A-body cars, however. The new 'Cuda was more expensive and heavier than the A-body cars, meaning that a Duster 340 could eat a 'Cuda 340 for lunch. Many buyers discovered the Duster 340, or Dodge Dart Swinger 340, were better performance values than the sportier-looking new ponycars.

Despite the Barracuda's humble A-body economy car origins, over the years it has become more closely linked to the Challenger, and is therefore not covered extensively in this volume. The tale of the Barracuda and later Challenger rates its own book. And the story of the Valiant, Duster, Dart, and Demon quadruplets can stand on its own as well.

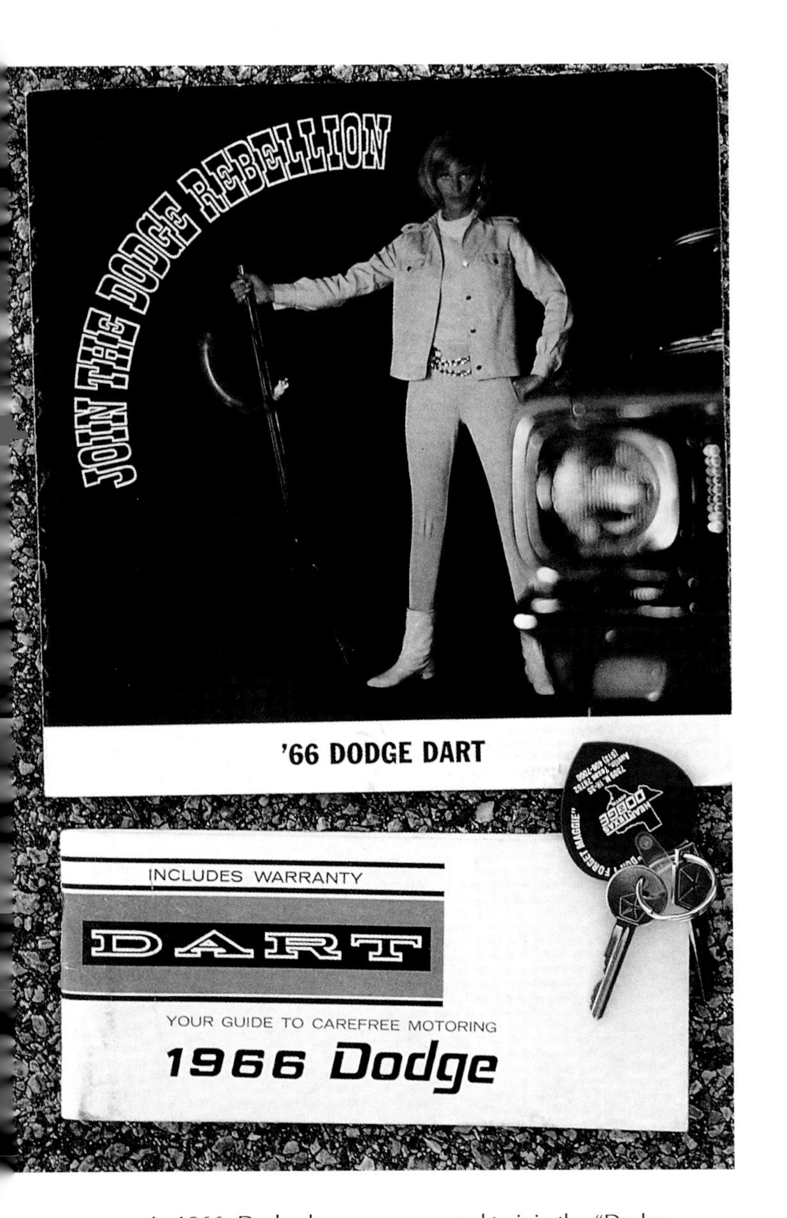

In 1966, Dodge buyers were urged to join the "Dodge Rebellion," with the marketing theme carried over into television, print ads, and even dealer brochures.

The 225, despite its common ancestry with the 170, exhibited a different personality from its smaller sibling. The 225 achieved its displacement from a very long stroke, and was the slower-revving of the two powerplants, although it produced more torque and horsepower. Like their Ford and Chevy rivals, the two late-arriving hardtop coupes from Chrysler came with bucket seats, special trim, unique emblems, and could be ordered with a more powerful engine than the stripped-down models.

Besides the compact Valiant and Lancer, Dodge had another entry in the low-price field, although one that would never be mistaken for a true economy car. Dodge introduced the Dart nameplate in 1960. The Dart was built on the regular Plymouth's 118-inch chassis (compared with the Valiant's 106.5-inch wheelbase), and as such was slightly smaller than the full-size Dodge Matador and Polara. Like the Valiant, the Dart featured unitized body construction.

The Dart came in Seneca, Pioneer, and Phoenix models. Although the Dart was sold with the 145-horsepower 225-ci Slant Six, buyers could also order 318-ci and 361-ci V-8s, the latter with dual four-barrel carburetors. The Phoenix represented the top of the Dart line; it came standard with the 318 V-8.

Rather than targeting the compact market, the original Dart was intended to battle the bread-and-butter Ford, Chevrolet, and Plymouth family sedans. The standard Dodge Matador and Polara competed more directly against the more stylish Pontiac and Buick. "These economy cars are designed to compete in price and size, series for series, with the automobile industry's low-priced Big Three," said Dodge's general manager, M. C. Patterson, at the car's introduction. "The Dodge Dart makes it the 'big four' now." The Dart's base price was roughly $500 less than that of the larger Dodges, and within a few dollars of its intended competition.

The D-500 option turned the Dart into a serious performance car. It included a ram-inducted, high-compression 361-ci V-8 rated at 320 horsepower. Chrysler's ram-induction system was a gasp-inducing sight—two Carter four-barrel carburetors perched atop long, curving intake runners. The carbs were spaced so far apart they nearly touched the inner fenders.

The Dart, along with the rest of the Dodge line, was restyled for 1961. Although the styling was cleaned up considerably, especially the grille and front bumper, sales slid. The D-500 option returned for 1961, although the 361 engine was slightly detuned. With a 9.0:1 compression ratio and four-barrel carburetor, horsepower rang in at 305. Dodge offered a Police Special version of the 383 four-barrel in the Dart, rated at 325 horsepower. The top Dart engine was the Ram Induction 383 rated at 330 horsepower. The D-500 badge had designated Dodge's performance package since 1956, but the name was retired after 1961. New go-fast packages were in the offing.

When the entire Dodge and Plymouth line-ups were downsized for 1962, the Dart was put on a diet as well, although the car was still larger than the compact Plymouth Valiant. Chrysler planners believed General Motors was about to make a major shift away

The top engine available in the 1966 Dart GT was the 273-ci four-barrel, and engine identification tags made sure everyone on the street knew it.

The A-body performance models from the 1960s were blessed with a higher level of detailing than the cars that were to follow in the 1970s. The 1966 Dart GT rewarded sharp eyes with details such as the stylized GT insignia on the fender moldings.

Deciphering The Code

Cutting through Chrysler Corporation's alphabet soup of body and engine designations can be daunting for those who haven't been living and breathing all things Mopar, but cracking the code is not as difficult as it first seems. (The Mopar name, by the way, came from the corporation's Motor Parts division. Although originally a name for the company's replacement parts and accessories, the Mopar tag has developed into street slang for any Chrysler Corporation vehicle.) Chrysler, like other automakers, gave internal designations to its various chassis configurations that would be used across various vehicle lines. As a matter of economic necessity many cars shared large numbers of parts, and were built off common platforms.

For practical purposes, try to remember the A-body, B-body, and C-body designations as small, medium, and large. The A-body cars are the focus of this book—the 1960–76 Plymouth Valiant, 1963–76 Dodge Dart, the 1964–69 Plymouth Barracuda, the 1970–76 Plymouth Duster, and 1971–72 Dodge Demon. The B-body cars were the intermediate models like the Charger, Coronet, Super Bee, Satellite, Belvedere, Road Runner, and GTX. The A-body and B-body cars made up the core of the corporation's musclecar contingent in the 1960s and 1970s. In today's collector car market, the B-body hot rods like the Charger R/T, Road Runner, GTX, and Super Bee usually command the highest prices, largely because they came exclusively with big-block engines. Often times the hottest small-block A-bodies were actually quicker than comparable B-body cars, but in today's heavily regulated market, a torquey big-block V-8 is a commodity that will never be available again and auction prices reflect that.

The C-body cars were the full-size 1960s-era monsters like the Dodge Monaco and Polara, late-1960s Plymouth Fury, and Chrysler 300 and Newport. To further complicate matters, the 1970–74 Dodge Challenger and Plymouth Barracuda had their own chassis designation, the E-body.

Engines likewise were given internal designations. The "A" V-8 engine family comprised the polyspherical head designs used from 1955–66. The wedge-head "LA" engine family, introduced in 1964, consists of the modern small-block V-8s such as the 273, 318, 340, and 360. In recent times however, with the polyspherical engine receding further back in the historical rearview mirror, the LA engines are sometimes referred to as A-engines. Although a handful of Darts and A-body Barracudas were built with big-block engines, most of the celebrated Mopar compacts earned their reputations with powerful versions of the LA engine family. The 273, 340, and 360 were all sold in performance trim, giving rise to the Dart GT, GTS, and Swinger 340, as well as the Duster 340, and 360.

The "B" and "RB" engine designations refer to the big-block V-8 family. The 361, 383, and 400 are B-engines. The larger displacement big-blocks, such as the 413, 426 Wedge, and 440, are labeled "RB" (Raised Block) since a raised block is necessary to accommodate a longer stroke. Then, of course, there's the 426 Hemi, but that's an engine in a class by itself. The B and RB big-blocks powered the 1967–69 Dart GTS 383, the late-1960s Hurst-built 440 Dart, certain late-1960s Formula S Barracudas and 'Cudas. The 426 Hemi was only offered in the 1968 Super Stock Hemi Dart and Barracuda.

from large cars, and they scrambled to adapt. What had once been full-size, 122-inch-wheelbase cars were shrunk down to fit on a 116-inch wheelbase.

As it turned out, General Motors had no intention of abandoning the large car market, and the car that started all the rumors turned out to be the intermediate-size 1964 Malibu. Chrysler Corporation's decision to downsize its Dodge and Plymouth family cars turned out to be way, way ahead of its time. The new sensibly sized Mopars had merit, but sales tumbled. The American buying public was all-too happy with its full-size automotive choices.

The trim 1962 Dart models, combined with the upscale Polara version, were expected to carry the entire "standard size" Dodge banner that year. After sales cratered, however, Dodge hastily raided the Chrysler stable for a larger body, and created the 880 series, which kept the year from being a complete catastrophe.

The 1962 Dart was identified by numbered designations. There was a base Dart, but also a Dart 330 and Dart 440 series. The 330 series provided midlevel trim, while the Dart 440 represented the top-of-the-line. The most important number was 413, however, which was the displacement of the top-option V-8.

Down in the economy ranks, Dodge's attempt to bend the Lancer in a sporty direction with a GT model was not successful. Bucket seats and a floor shifter do not necessarily transform an economy car into a genuine GT (Grand Touring) automobile.

In 1963 Dodge eliminated the Lancer name (until it was applied to a front-wheel drive model in the 1980s) and affixed the popular Dart nameplate to a new economy sedan. That year Plymouth also restyled the Valiant, and the A-body train was set to roll in a performance direction.

Lancer Dies; GT Lives

Although intended to occupy the economy rung on the Dodge ladder, the 1963 Dart was a larger car than the Lancer it replaced. The Dart even rode on a 5-inch longer wheelbase than its corporate stablemate, the Valiant. In part this was in keeping with Dodge's position in the corporate hierarchy. The Dart's extra size meant the car could offer more room and a smoother ride than the entry-level Plymouth Valiant, not to mention other competing makes. In fact, Darts were later marketed as being the largest economy car one could buy. If that seems contradictory, welcome to the world of automotive marketing.

The 1963 Dart was another Virgil Exner design, and a considerably cleaner one than the earlier Lancer. The body had fewer of the exaggerated character lines that had typified Chrysler products of the late 1950s and early 1960s. The Dart's concave grille and single headlamps gave the car a distinctive, yet simple face. The Dart was available as a two-door or four-door sedan, and a convertible model moved the car a few more notches in a sporty direction.

The closest a buyer of a 1963 Dart could get to an actual performance car was the GT model, although at this stage of the Dart's development it was hardly the car to strike fear into the hearts of drag racers. The top engine was still the 225-ci Slant Six. The GT model provided the top-line upholstery, a padded dash, full-wheel hubcaps, GT trim, and bucket seats.

The changes worked wonders for Dodge's sales, however, as 1963 Dart production more than doubled over that of the 1962 Lancer. Of the 153,922 Darts built, some 34,300 were GT models.

The 1963 Valiant was also reskinned and revamped. It, too, lost most of its quirky styling. The profile was crisp, the character lines far more modest. The dual headlamps on each side were replaced by single units.

As the corporation's price leader, the Valiant was even more economical than the Dart. Valiant pricing began roughly $70 below the Dart's price tag, and weight was 100 pounds less. The Valiant's savings came from its smaller size. It rode on a short 106-inch wheelbase, and was 9 inches shorter overall. Plymouth kept the V-100 and V-200 designations for the entry-level and midlevel models, respectively.

The top Valiant continued to be the Signet model. Like the Dart GT, the Signet's main selling features were its bucket seats, full wheel covers, extra interior appointments, and Signet trim. The 225-ci Slant Six was an option over the base 170 Six. The 225 was impressive for its class, but still no performance engine.

That all changed for 1964. The upward migration of the American economy car had been in progress almost since the beginning, and Chrysler Corporation went with the trend. Chevrolet decided to supplement the Corvair with a larger, more conventional Chevy II in 1962, and for 1964 the newcomer could be had with a 283-ci V-8. Ford had slipped its 260-ci V-8 under Falcon hoods starting in 1963. The Studebaker Lark could be ordered with a supercharged V-8. For 1964 the Dart and Valiant got their V-8s, and for the first time there was real performance potential to be tapped in the corporation's compacts.

Prior to 1964, Chrysler Corporation's "small" V-8 was the 318-ci A-engine, with polyspherical cylinder heads. The polyspherical 318 was a physically large, heavy engine, especially for the displacement it offered. In fact, it was too large to fit comfortably in the nose of an A-body Mopar.

The new V-8 addressed those shortcomings. The 273-ci V-8 that debuted for the 1964 year model was considerably lighter and more compact than the polyspherical 318. Throughout the early 1960s advances in casting technology had allowed all automakers to replace their heavy, bulky engine designs with powerplants that utilized stronger, more compact engine blocks. The new generation engines also incorporated trimmer cylinder heads and lighter valve train components for further weight savings.

Chrysler's new "LA" small-block V-8 was one of these slimmed-down engines. But it wasn't completely a clean-sheet design—in many ways it was more an offshoot of the polyspherical 318 than an all-new engine. The two V-8s shared a number of parts, including the crankshaft, connecting rods, and bearings.

But the LA V-8 brought some new advances to the fore, most noticeably the thinwall block casting, which saved roughly 40 pounds over the older design. Chrysler also dumped the polyspherical head design in favor of cylinder heads with wedge-shaped combustion chambers, just as the big-block 383/413 wedge had replaced the 1950s-era 354/392 Hemi-head V-8s.

The new engine displaced 273 cubic inches and carried an 8.8:1 compression ratio. Fitted with a Carter two-barrel carburetor, the 273 was rated at 180 horsepower at 4,200 rpm and 260 ft-lb of torque at 1,600 rpm. It was available in the Dart, Valiant, and Barracuda, and provided good, though not breathtaking performance.

"To be known as the Charger 273, this engine will be the smallest and lightest V-8 built by Dodge but will offer a performance potential equal to that of the [polyspherical] 318-cubic-inch engine now provided in the standard-size Dodge V-8 models,"

GT insignia on the interior was minimal on the 1966 models, with the inside door emblem being the most prominent. A remote outside mirror control was optional.

said Dodge general manager Byron Nichols upon the engine's introduction.

Advertising for the 1964 Dart touted the V-8's capabilities by tying it to the popular Slant Six. "Performance lovers are pretty much in agreement about Dart's optional 225 cu. in. Six engine. They say it has the muscle of an eight. If you check the facts and figures, you'll have to agree with them: It's a tough engine to beat," One ad boasted.

"But we've got news for you. You can. Now there's a new engine in the Dart stable . . . a hot new V-8 that's 273 cubic inches strong. This new V-8 was especially built for Dart. And that's as it should be."

The V-8 was a good fit with the compact A-bodies, but most buyers weren't convinced, at least initially. Most A-bodies were still ordered with one of the available six-cylinder engines. The cars were moving up in the world, but were still sought primarily for economical transportation.

The Dart and Valiant also received some modest styling changes for 1964. The Dart's concave grille was replaced by a more conventional design. The GT was given a fat beltline molding. The Valiant was likewise given a new grille, although the big news from the Plymouth side was the Barracuda (see sidebar).

Dodge continued its strategy of positioning the Dart as the largest of the compact cars. The 1964 dealer brochure pictured a Norman Rockwellesque illustration of two boys, complete with slingshots and baseball gloves, with a lead caption of "My dad's compact is bigger than your dad's compact!" The text advised that "Boys will be boys. And, as in the past, compacts will be compacts—with one handsome exception, pictured above. That's Dodge Dart, a fresh, new compact in the large economy size."

Small Muscles

The 273 V-8 began fulfilling its potential in 1965. That year saw the debut of a genuine high-performance version of the engine. The hot rod version of the small-block, called the Commando 273 at Plymouth, Charger 273 at Dodge—turned Chrysler's compacts into pocket rockets.

Chrysler used a conventional hot rodding path to bulk up the 273. The engine was given a Carter AFB four-barrel carb, naturally. The engine's compression ratio was upped to 10.5:1, and the hydraulic lifters were replaced by solid lifters and cam. The free-flowing single exhaust provided a noticeable growl, so hiding the engine's intent was difficult. The bolstered 273 was visually impressive as well. It was dressed with crinkle-finish valve covers and a large chrome air cleaner with either "Commando 273" or "Charger 273" identification. The high-performance 273 was rated at 235 horsepower at 5,200 rpm, and 280 ft-lb of torque at 4,000 rpm.

With the added power came extra demands on the rest of the powertrain, so Dodge added a Heavy-Duty (HD) Sure-Grip 8 3/4-inch rear axle option to the Dart's tech sheet in 1965. The HD 8 3/4 rear end came with a 4.89:1 gear ratio. Offset rear wheels had to be used with the beefy rear end, although that wasn't much of a problem for the racers for whom the option was intended.

Hot Rod magazine put a 273 Commando engine, as tested in a Barracuda, through the grinder. "In all passing situations I encountered, of which there are many on a trip of this character, an abundant power reserve was always on tap," wrote Eric Dahlquist.

"Only once on a long, lonely road did I actually let the car out to over 115 miles per hour indicated and the speedometer was still climbing before backing off. Larger machines and expensive imports were always surprised to see a 'compact' passing, without, as the exhaust note revealed, apparent strain."

Musclecar engines of the 1960s were not noted for fuel economy, but the *Hot Rod* testers averaged a very good 15.5 miles per gallon. In the quarter-mile they clocked in at 16.46 seconds and 89 miles per hour.

The 273 Charger engine gave the Dart GT true "Grand Touring" credentials. Other changes helped push the Dart and Valiant in a more mainstream direction. The push-button automatic transmission controls were dropped from all Chrysler products in 1965. The Dart and Valiant were both given new sheet metal that cleaned up styling even more. Both cars received new grilles, bumpers, and hoods.

Oddly, Chrysler gave the Dart a mild sheet metal facelift for 1966, despite the fact that thoroughly restyled Darts and Valiants were on the drawing boards for 1967. The Dart was given new fenders with stamped character lines at the leading edge, and the yearly grille upgrade. In the interior, the optional center console was upgraded to the more substantial, full-length type used on Dodge's upscale cars. The console would become a recognizable feature of Mopar performance cars for years to come. The seats and interior door panels were restyled, and new fabric patterns introduced.

The Dart continued to be marketed as a "big" compact. Dodge's new ad campaign for 1966 centered around the "Dodge Rebellion" theme, as the company tried to tap into the growing youth movement. It was a ridiculous ploy, implying that buying a Dodge was somehow an act of rebellion, but marketers were scrambling to somehow capitalize on the growing 1960s hairball of campus protests, drug culture, sexual liberation, and hippie-inspired fashion.

The self-described "Dodge Rebellion gal" opened Dart television commercials by asking, "Do you own one of those itty-bitty cars that makes you feel like you're driving in solitary confinement?" A narrator drove the point home by admonishing listeners to "Help stamp out cramped compacts." The Dart was described as "The compact with impact," that had "more power than you can shake a stick at." Print ads advised potential buyers to "Avoid the cramped compact squeeze" and purchase a "man-sized" Dart.

Mechanical changes for the 1966 A-bodies were incremental, but worthwhile. The cars received thicker brake linings, a new power steering pump, and a new universal joint design to reduce vibration. Front disc brakes were optional. The four-speed shift linkage was revised, and a reverse-lockout feature added, although most Mopar enthusiasts consider the switch to this new Chrysler/Inland shifter a step backward.

An AM/FM radio was a dealer-installed option in 1966.

The high-performance 273 was still the top engine option, but the spotlight had definitely shifted to other cars in the Chrysler line-up. The Dodge Charger made its debut, complete with radical fastback sheet metal and the new street version of the 426 Hemi. The Dodge Coronet, Plymouth Belvedere and Satellite were the other recipients of the monster motor. Suddenly, 273 cubic inches didn't seem like so much anymore. Soon enough, some of those cubic inches would trickle down to the A-bodies.

GTS
383

two

Middleweight Contenders

Dart GTS and Swinger, 1967–1969

In 1967 the Dart and Valiant grew up—both literally and figuratively. The two cars were physically larger and more grown-up in appearance. The sheet metal no longer dragged protesting eyeballs along exaggerated body lines the way the first Valiants had. The new 1967 bodies were tasteful and cleanly designed, just like the big cars.

The Chrysler A-body cars were not alone in this styling shift. All of the Big Three automakers were cleaning up their designs and de-chintzing their economy cars wherever possible. Both the Chevy II and Corvair had evolved into handsome small sedans. Ford's Falcon was perhaps less successful in this regard, but there was still a big difference between the frumpy 1960 Falcon and a 1966 model. The new small Chryslers were right in step with the times. And Elwood Engel, Chrysler Corporation vice president of styling, had always liked the squared-off look anyway.

The Dart styling team was led by Carl Cameron. The new car featured a recessed backlight fitted between modest sail panels that looked good with

The 1968 Dart GTS adopted the trademark "bumble bee" stripes that identified high-performance Dodges, although buyers could choose an optional stripe that ran along the car's flanks from nose to tail or even check the "stripe delete" box on the order form. Other new styling cues included the small round marker lamps, a new grille, and the aggressive-looking, vented hood. A large GTS emblem replaced the Dodge lettering on the leading edge of the hood. The car pictured features Cragar S/S mags, a popular choice for cars of the period.

The 1967 Dart's new sheet metal was clean and crisply folded, giving the Dart a "big car" look, despite the new model being almost identical in size to the 1966 version. Also adding to the "big" compact image was a new 383 big-block engine option.

the hardtop's handsome roofline. The character lines on the side of the car were well–proportioned. The hood was mostly flat, but a drop-off at the leading edge gave the front end some personality. The station wagon model was dropped from both the Dart and Valiant lines, leaving Dodge with a two-door hardtop, two-door sedan, four-door sedan, and two-door convertible.

The 1967 Plymouth Valiant shared the same cowl and windshield with the Dart, plus the front doors. Beyond that the Valiant had its own styling signature, although the overall shape was very close to the Dart's. The Valiant's hood featured a character line down the center, and the character line that ran down the side of the car kicked downward right before the rear wheels. The Valiant also had its own distinct grille and rear end treatments.

"The Valiant for 1967 could very well be the sleeper of the year—and that's speaking from the styling viewpoint alone," reported *Popular Mechanics* writers in the October 1966 new car preview. "Entirely retooled sheet metal from roof to rocker

The 383 V-8 was such a tight fit in the Dart's engine compartment that power steering was not available. The 383's driver-side motor mount occupied the space that the power steering pump normally claimed. Even with the manual steering, owners noticed that steering linkage tended to rub against the engine's oil pan. The exhaust manifolds were so restrictive that horsepower was down from the normal 325 to 280.

Chrome valve covers and "pie pan" engine identification were standard features of the 383 GTS. A dual-circuit master cylinder was a new, and welcome, feature of the 1967 Darts.

panel has resulted in a body line that's got more class than you could hope to expect in exchange for so few dollars."

The Valiant was marketed as part of the "Plymouth is out to win you over," ad campaign, with television commercials promising the new Valiant would be "As easy on your pocket book as it is on your eyes." Playing up further on the Dart/Valiant theme of being the "big" compacts, TV ads claimed "You'll have to keep reminding yourself it's a compact." In print ads, potential buyers were urged to "Revolt against kiddy car compacts. Go '67 Dart."

Since the Barracuda was carrying the A-body performance banner for Plymouth, the Valiant's body-style choices were on the stodgy side. Plymouth dropped the convertible and two-door hardtop from the Valiant line. Two-door and four-door sedans were all that survived.

Shopping for a Valiant was still pretty simple. The two available models were the stripped-down V-100 and the slightly less barren Signet. The Signet included substantially more bright trim than the V-100, along with a light group for the trunk, glove box and map reading; extra padding for the dash and sun visors; fender-mounted turn signals; and the larger 225-ci Slant Six was standard. Although the Valiant was not destined to share in the horsepower bounty that was right around the corner, at least the base 170-ci Slant Six got a boost from 101 to 115 horsepower, thanks largely to a camshaft change. The base price for a V-100 two-door was a rock-bottom $2,117.

The A-bodies, besides growing on the outside, also grew up in the engine compartment. In 1966 the 273-ci V-8 had been the largest available powerplant for either car. For 1967, the 273 two-barrel and four-barrel were still around, but the trend in American cars was toward ever larger engines. The horsepower war was accelerating, and even the compacts and ponycars were foot soldiers in the war. Ford had widened the Mustang's flanks just enough to fit its 390-ci big-block underhood for 1967, and Chevrolet's Camaro offered a 396 shortly after the car's introduction. Chevy also had a 396 Nova SS in the

A four-speed transmission was standard with the 383 Dart GTS, with the TorqueFlite automatic optional. The wood-trimmed steering wheel from the elusive plastic tree was optional.

The 1967 Dart interior used painted metal upper sections on the doors with vinyl-covered lower sections. GT models had their own identification.

works for 1968. Plymouth kept the Barracuda right in step with these rivals, wedging the 383 big-block between its fenders.

Fred Schimmel, a stylist in the Plymouth studio from 1955 to 1975, remembered that the corporation had been working toward a big-block compact for some time. "They did some experiments when I was there, they stuffed a 383 in a Valiant and they let us drive it one time," he recalled. "Well, it crowded the engine compartment so much that they didn't have room for power steering, and it was so heavy it took two people to turn the steering wheel."

The production version wasn't much different. In Plymouth's haste to stay abreast of the competition it took what could charitably be described as shortcuts to get the largest engine possible in the A-bodies. If not actual shortcuts, they were at least deviations from the norm. The Formula S 383 Barracuda could not be ordered with power steering, although there was room for air conditioning. The 383's exhaust manifolds had to be redesigned to fit, resulting in a substantial power loss. The 1967 Barracuda's 383

Only 457 Darts were built with the GTS package in 1967, a fraction of which were convertibles. The low number was due in part to the GTS' late introduction. GTS production reached 8,745 in 1968.

The first-year Dart GTS carried GTS 383 emblems on the fenders, but still had GT badges on the grille, trunk panel, and inner door panels.

was rated at 280 horsepower and 400 ft-lb of torque, compared to 325 horsepower for the regular 383 four-barrel available in other models.

Still, Eric Dahlquist, writing for a *Hot Rod* magazine road test, was impressed with the package, although he, too, couldn't help but notice the tight fit. "Because of the close proximity of the exhaust manifolds to the inner fender panels, we never expected that plug maintenance would be easy, but we're here to tell you that the job's almost suicidal when the engine's hot," he noted. "In the first place, the task requires patience, and in the second, your hands will be candidates for the First Aid Cream award."

Fortunately, they managed to keep their tender skin intact long enough to record a 15.15-second quarter-mile at 92 miles per hour. While respectable, the 383 Barracuda was not going to knock Plymouth's own GTX, or Dodge's Hemi Charger, from the top of their perches.

Late in the model year, Dodge released its own version of the big-block compact, although the division required some prodding. Brothers Norm and Leonard Kraus, owners of Grand Spaulding Dodge in Chicago, had built a reputation for their dealership selling a racing image along with high-performance tuning services. They were especially hot for a big-block Dart, and decided to do something about it.

"We saw that all we had was the Super Bees, and the Coronets, and all those bodies," Norm Kraus recalled. "Those were heavy cars—they were performance cars but they were heavy cars. We didn't have any image. When we talked to the factory, they said, 'Hey, we're going to come out with a smaller car.'

"I said, 'Yeah, what's that?'

"They said, 'Well, we're going to call it the Challenger. We're working on that now.'

"I said, 'Well, is that going to be like the size of a Corvette?'

"They said, 'Yeah, just the same size as a Corvette.'"

Kraus came away convinced the Challenger was promising, but the E-body Challenger was still a couple years away. "What do I have for today? I have the same cars," he remembered.

"They kept saying the big-block V-8 would never fit. So I called over to my parts guys and my service guys, and said, 'Hey, let's get a Dart in here. Let's see what we can do. They say they can't get the big-block V-8 in the Dart. Let's see what the problem is,'" Kraus said.

One reason Mr. Norm was so keen on dropping a 383 in a Dart was that the smaller 273-ci V-8 wasn't cutting it, in his eyes. "We didn't think too much of them," Kraus said of the 273-ci Dart GTs. "You can't capture the imagination with a 273 Dart, when the Camaros are running 350s."

The Dodge Daroo I hit the show car circuit in 1968. With its cut-down roof and glass, pointed nose, and velocity stacks, it was hard to find the Dart underneath all the swoopy bodywork. But traces of the stock Dart dashboard could be seen among the Daroo I's new interior features. *Reprinted with the permission of the DaimlerChrysler Corporate Historical Collection*

Low-back bucket seats in the Dart GT and GTS were cleanly styled, but the small metal emblems embedded in the seat backs doubled as branding irons on hot days.

They quickly pulled the engine from a lesser Dart, dropped in a 383, and went motoring. "All we had to do was trim about a half-inch off the K-frame in the front," Kraus said. "We had to make a little change on the left-side motor mount, and it fit in there without any problems. So I said, 'The exhaust is a little close to the steering coupler. Why don't we put a heat deflector on there?'"

They did, and found the 383 worked fine in the newly restyled Dart. "We started driving it around, the car handled beautifully," Kraus said. "We were thinking of image . . . advertising.

"So we called the factory at that point, and said, 'We've got a 383 in a Dart. We'd like to bring it down and show it to you.' So [Robert] McCurry, who was the head of the Dodge Division at the time, said, 'Bring it down.'

"We drove into Detroit, and we brought the car right into the plant. We got into the plant, he comes downstairs, looked at it, he got in the car and drove it around, and says, 'Hmm.'

"He goes to the telephone and calls the Engineering Department and says, 'You guys better come down here. I want to show you what these two kids from Chicago did,'" Kraus recalled.

"Two engineers come down—with pipes, just like you'd expect it—they came over and took a look and said, 'Well, this has got a heat deflector on it!' So

McCurry says, 'What's the matter, couldn't you do that?'"

After making the case to Dodge management, and getting a positive response, Grand Spaulding Dodge started advertising the availability of 383-ci Darts immediately, Kraus said. The impression left among Mopar enthusiasts was that Grand Spaulding was building their own unique 383 Darts. It was an impression that worked well for the performance-oriented dealership, even if it did put the factory in an uncomfortable position.

"They said, 'Look, if we build a car here we have to build it for everyone,'" Kraus recalled.

"We said, 'This is our car. How can we get around this?'

"He says, 'We can do this, but you've got to order 50 at a time. And the other dealers have to order 50 at a time.' So I said, 'Fine, you've got an order for 50,'" Kraus said.

"That's how it started. Then we started using all the information. 'Now the Dart comes with a 383, the total performance package.' We just threw all the advertising into the areas where we were—the radio, TV, and the tracks, the performance papers, *Drag News, Car Craft* magazine. At the same time, all the magazines came in to do stories on them.

"It was a completely stock engine. I think it was about 800 pounds less than a Super Bee with the same engine," Kraus said. "We could live with that."

Even in 1968, long before such safety features saturated the market, Chrysler offered childproof door lock knobs. Few ordered the feature (especially for use on big-block Darts), but such ideas now infest new cars from stem to stern.

Dodge built a separate option around the 383 engine, and named the finished product the Dart GTS. The GTS package could be ordered with either the Dart GT hardtop or convertible. To help his dyno-tuned big-block Darts stand out from the crowd, Kraus removed the "T" from the factory GTS emblems, and substituted an "S," creating the GSS, or Grand Spaulding Sport. That started a yearly run of GSS Dodges that ran through the 1972 model year.

To keep the 383 on a leash, the GTS package came with a heavy-duty suspension package of big-block-spec torsion bars, rear springs, idler arm, pitman arm, and A-arms. Disc brakes were included, as were redline wide-oval tires, and a four-speed transmission, standard. A TorqueFlite automatic was optional.

Chunky GTS emblems were located on the trunk, fenders, and hood of the 1968 models, plus on the inside door panels. The 383 cars received engine identification, the base GTS did without.

As can be expected, the Dart GTS 383 also came standard with the same shortcomings as the A-body Barracuda. Horsepower was down to 280 and maintenance was iffy thanks to the restrictive exhaust manifolds. Buyers could look forward to serious upper body workouts, as they were not going to get an assist from power steering.

However, the 383 Barracuda and Dart GTS made sense to those whose primary interest was racing, whether legally at the strip or illegally on the street. The 383 A-body offered a lot of potential for modification, or at the very least, a lot of potential for tire-smoking giggles on the street.

Another Year, More Performance

For 1968, most of the action was on the small-block side of the engine family. The 273 small-block and 383 big-block V-8s worked well enough for their intended purposes, but there was plenty of room for something in the middle that could bridge the gap between underpowered small V-8s and unwieldy big-blocks.

In 1968 the A-bodies got two such somethings.

The first was a 318-ci version of the LA small-block family. The new 318 had actually been introduced in 1967, but only in larger Dodge and Plymouth models. Chrysler Corporation had had an earlier V-8 with 318 cubic inches of displacement, the polyspherical-head V-8 used in the late 1950s and early 1960s, which caused some confusion. And in fact, the new 318 shared much with the old, just as the 273 did. But the new smaller, lighter, 318 made a lot of sense for the A-body cars. It was a good midlevel engine, offering more power than the 180-horsepower 273 without inflicting a weight penalty. With its two-barrel carburetor and single exhaust, the 318 was rated at 230 horsepower. That well-rounded nature allowed the 318 to become a staple of the Chrysler Corporation for three decades, powering virtually every car in the line-up at some point, as well as the pickup trucks.

The other newcomer for 1968 was perhaps less well rounded, but certainly more welcome to lovers of fast cars. The 340-ci LA-series small-block made its debut in the Barracuda and Dart GTS and offered just the right combination—midsize displacement, very good power, and compact dimensions.

The 340 was intended strictly as a street performance engine; Chrysler never built any two-barrel carb, single-exhaust versions for family station wagon duty. As a member of the LA engine family though, the 340 shared much with its 273 and 318-ci cousins. The 340 engine utilized the same stroke as the 273 and 318, gaining its extra displacement through a larger 4.04-inch bore. For strength, the 340 was fitted with a forged steel crankshaft,

The Daroo II Dart show car turned heads with reproportioned bodywork that lengthened the hood, shortened the rear deck, and chopped the top. Dodge described the results as a "close coupled competition roadster," in press material. Powered by a 340-ci V-8, the Daroo II was an operable car. *Reprinted with the permission of the DaimlerChrysler Corporate Historical Collection*

windage tray, and double roller timing gears and chain. As was typical of performance engines, the 340 squeezed out a 10.5:1 compression ratio.

With 2.02-inch intake valves and 1.60 exhaust valves, and larger ports than those used on the smaller engines, the heads offered good breathing capabilities. A Carter AVS four-barrel carburetor and dual-plane intake manifold passed air and fuel through those heads. The exhaust manifolds were shaped with heavy breathing in mind as well. A dual-point distributor and high-voltage coil sparked the fuel mixture.

Those first 1968 340s were even available in two distinct performance flavors. The 340s destined for four-speed cars came with a hotter cam that featured both more lift and duration than those 340s teamed with a TorqueFlite automatic. Unlike the high-performance 273, though, the 340 V-8 relied on hydraulic lifters, a more streetable combination than the more radical solid-lifter and cam setup. The 340 was rated at 275 horsepower, but actually put out closer to 310 or 320, by most accounts.

With a 275-horsepower small-block available, 1967's 280-horse 383 would have been, at the very least, a redundancy for the new model year. To keep the 383 as the legitimate top engine, Dodge engineers addressed some of the problems that had made the engine an underachiever in 1967—namely, limited breathing capabilities. The 1968 Dart GTS and Barracuda 383 were given new cylinder heads with larger valves and 10 percent more intake port area. Chrysler also hogged out the intake manifold, achieving 10 percent greater branch and 50 percent greater runner sizes. Along with changes to the vacuum advance in the distributor, the fiddling resulted in a bump to 300 horsepower, although the exhaust manifolds were still the cork in the horsepower bottle. (Technically, the 383 wasn't even the senior engine in the 1968 A-bodies. Chrysler offered the ferocious 426 Hemi in the Dart and Barracuda, a subject covered in greater detail in chapter 3.)

The 340 supplanted the 383 as the standard GTS engine. In 1968 standard GTS equipment included

Chrysler's "LA" small-block engine family progressed from the 273-ci in 1964, to the 318-ci in 1967, and to the powerful 340-ci V-8 in 1968. Designed strictly as a performance engine, the 340 was never even offered with a two-barrel carburetor. (Well, technically it was sold at one point with a trio of two-barrel carbs . . .) The 1968 340 came with a 10.5:1 compression ratio, and was rated at 275 horsepower at 5,000 rpm. In 1968, 340-ci engines teamed with manual transmissions received more aggressive camshafts than those bolted to automatics. Cutaway reveals the engine's strengths—double roller timing chain, forged steel crank, and oil pan windage tray. *Reprinted with the permission of the DaimlerChrysler Corporate Historical Collection*

Horsepower was up from 280 to 300 in the 1968 GTS 383, thanks to revised cylinder heads with larger ports and exhaust valves. The 1968 edition also used a smaller air cleaner assembly than the 1967 version.

The 340-ci V-8 was a much better fit in the A-body cars than the hulking big-block 383, and even offered superior real-world performance. In 1968 and 1969, the 340 was available in the Dart and Barracuda. The Valiant line offered no performance version in the late 1960s. *Reprinted with the permission of the DaimlerChrysler Corporate Historical Collection*

the 340 V-8 (383 optional); standard four-speed transmission (TorqueFlite automatic optional); Rallye suspension and heavy-duty shock absorbers; E70x14 redline tires; a new-for–1968 hood with nonfunctional "power bulge" scoops; bumble bee stripes; and GTS identification on the hood, deck lid, and fenders.

Styling and interior changes for 1968 were not extensive, but were effective. A new grille for both the base models and GTS incorporated round parking lamps. The GT and GTS grilles were partially blacked-out, with minimal bright highlights. The rear trunk panel was given a new appliqué. A new bodyside molding was both attractive and functional.

The GT mimicked the GTS in many ways, but did without the power bulge hood and bumble bee stripes. GT engine choices started with the 170-ci Slant Six, with the 225-ci Six, 273-ci two-barrel V-8 or 318-ci V-8 optional. Cars ordered with the V-8 were fitted with small "V-8" emblems on the fenders. The high-performance 273-ci four-barrel V-8 had been canceled after the 1967 model year.

The Dart GT model continued for 1969. Although the sheet metal remained the same, the Dart was given the usual cosmetic makeover for the new year. Rectangular parking lamps replaced the round units used in 1968. The GT received a body-length belt-line molding and Dart GT emblems on the quarter panels. The GTS was replaced by the Swinger 340 model in 1970. A GT option was listed as being available on 1970 Dart Customs, although so few were built most consider the 1969 GT the end of the line. *Reprinted with the permission of the DaimlerChrysler Corporate Historical Collection*

In the interior, an additional cushion was added to the lower edge of the dash. The new cushion's job was to protect occupants' knees in case of collision, but most Dart owners found the feature more useful as a repository for spare change or bottle caps.

With the addition of the 340 V-8 to the mix, the GTS had clearly come into its own. Although an economy car at heart, the GTS had performance equal to that of many larger, more expensive musclecars. Handling was good, and the 1968 GTS featured the popular Mopar musclecar styling enhancements of the day, such as hood "power bulges" and bumble bee stripes.

Just as the GTS was establishing itself, though, a new breed of musclecar was being hatched at Plymouth. Hoping to collect the young buyers who were being priced out of the musclecar market by escalating window stickers and high insurance costs, the division released the Road Runner. Based on the intermediate-sized Belvedere, the Road Runner combined stripped-down performance with a cartoon-influenced image to great effect. The Road Runner came with just the basics—383 four-barrel, four-speed, bench seat, plain hubcaps, and just enough musclecar styling cues to get the point across.

The Road Runner left the GTS in the unusual market position of being a smaller car with a smaller engine, but with a higher base price ($3,163) than the Road Runner ($2,913). Of course, many buyers went ahead and loaded up the Road Runner with its many available options, largely defeating the purpose of the car, but its image was established quickly. And the Road Runner's success would influence the evolution of future Dart performance models.

The Dart GTS did well in its own little niche, though. As *Hot Rod* magazine reviewer Steve Kelly

The later 1968 and 1969 Dart GTS models offered a unique Hurst shifter for the four-speed transmission. The unusual curved and bent lever placed the shifter closer to the driver and further away from knuckle-busting encounters with the dash. Although strange looking, the bent Hurst was a big improvement over the Inland shifters used from 1966 to midway through 1968.

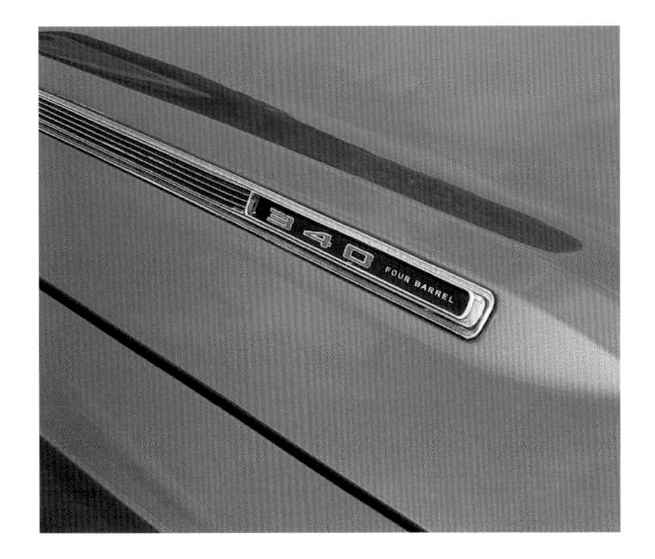

A 275-horsepower 340 V-8 resides under the hood of this 1969 GTS.

The 1969 Darts traded the previous year's round side marker and parking lamps for rectangular units. The only way to get a Dart convertible in 1969 was to order the GT or GTS model. The last Dart convertibles were the 1969 models. Both horsepower and sun worshipping were nearing their peak. By the early 1970s, convertibles and big V-8s were on their way out.

Hood vents on the 1969 GTS held engine identification, for either the 383 or 340-ci V-8s.

noted, "GTS could mean a lot of things (careful, watch those imaginations), but it's what it does that counts. It performs rapidly going straight or around corners, it stops quick and straight, and carries a reasonable price. It also outperforms cars with [an additional] 50 or more cubic inches of displacement."

Hot Rod's best quarter-mile elapsed time (ET) was a 14.38-second pass at 97 miles per hour (with the air cleaner lid removed), and the magazine testers ran easy 14.5s in completely stock trim.

"Not to take the edge off the Road Runner, the GTS might be a more sensible package. The base price is higher, but you get things like carpets on the floor, fat tires, bucket seats, and a few other niceties that can make Saturday night roaming more comfortable," Kelly wrote. "The GTS swings in many ways. It's got plenty of room for rear seat passengers and will serve bachelor and family man with equal aplomb. No need for the young guy to spend bucks making his family wagon look like it isn't. The GTS isn't going to give him away. It'll deliver super performance for 'weekend warrioring' and stand the thrashing without complaint."

The simple GTS tag of previous years morphed into GT Sport in 1969. Although the 1969 edition still carried a GTS emblem on the hood, the bumble bee stripes and door trim spelled it out, "Sport."

Car and Driver magazine recorded a 14.4 quarter-mile at 99 miles per hour in its test of the GTS in the September 1968 issue. The editors were amazed at the 340's capabilities, noting "...the intrepid Chrysler engineers bravely stuck their necks way out and rated this strong-hearted little engine at 275 horsepower at 5,000 rpm. We'd be the last to accuse anyone of underrating but the underground isn't kidding when they say 340s shoot Darts down the road in a 350-horsepower fashion." They pronounced the 340 as "easily the most exciting engine Chrysler has produced since the Hemi."

Car and Driver also zeroed in on one of the GTS' more unusual features, the wildly bent Hurst shifter. The Hurst, introduced midway through 1968, was more than just for looks—it was a vast improvement over the previous years' Inland shifter. "The stylists, with the stylists' eye for what's right, decided that it must come through the widest part of the console and right in the middle," *Car and Driver* noted. "That the hole in the console is a foot forward of the spot where the shifter comes through the floor is merely a problem for the engineering department. The result is a shift

The 1969 Dart GTS stood with two tires in the economy car world and two tires in the musclecar world. The 340 V-8, hood bulges, and bumble bee stripes were pure 1960s supercar, while the simple ornamentation and dog-dish hubcaps were right at home in the compact car class. With its 340 V-8, however, the Dart GTS was never overlooked on cruise night.

One extremely rare option for 1969 model Mopars was the W23-code cast aluminum road wheel. These wheels had barely been introduced before the factory recalled them, due to problems with the lug nuts coming loose. Almost all of the road wheels were replaced at the factory or by dealers before the public even saw them. Surviving examples, especially the few manufactured for the A-body's smaller bolt pattern, are prized by collectors today.

lever that looks as if Charles Atlas got mad at it, but fortunately function is in no way impaired."

The Hurst shifter wasn't the only deviation from standard practice for 1968. In a much wilder vein, Dodge released the "Daroo I" and "Daroo II" Dart concept cars on the auto show circuit. The Daroo I was almost unrecognizable as a Dart. Its pointed, shark-like nose, induction trumpets bursting through the hood, sidepipes, and cut-down windshield looked more like the work of a drunken Hot Wheels toy car designer than something that would ever sit in a Dodge showroom.

The Daroo II's most prominent features were its reproportioned shape and low-cut silhouette. Starting with a GTS, the designers chopped the car's height from the stock 52.8 inches to 42 inches high.
continued on page 53

The 1969 Darts were given a new black aluminum rear panel treatment to go with the revised grille. The GTS' rear bumble bee stripe was a delete option. *Photo courtesy Rob Reaser*

Moving Iron—Dodge Scat Pack and Plymouth Rapid Transit System

It's sometimes called a "halo effect," but it can also be described as the automotive equivalent of a share-the-wealth philosophy. That is, if a company has a high-profile car with a good reputation, it's considered good marketing to spread that reputation throughout the car line like butter on hot biscuits. Thus, the top-line Ford V-8s in the late 1950s and early 1960s were called "Thunderbird V-8s," and in more modern times, "Ram tough" trucks begat ram's head emblems on Dodge cars.

Tying disparate cars together around a common theme is a sure way to get a message across. Dodge and Plymouth embraced this philosophy aggressively in the 1960s and 1970s with their performance cars. The hot Dodges had a few standout features that the division bundled together for advertising purposes. Starting in 1968, the Dodge musclecars came standard with "bumble bee" stripes around the tail. The markings derived the name bumble bee stripes from Dodge's Super Bee, the low-cost alternative to Plymouth's Road Runner. The big story of 1968, of course, was the handsomely restyled Charger, which had earned good press reviews and outsold the 1967 Charger several times over. Dodge molded the plush Charger, cheapo Super Bee, and compact Dart GTS together into the Dodge Scat Pack.

The Scat Pack ads began appearing in 1968, with the bumble bee-striped Charger R/T, Super Bee, and Dart GTS receiving the attention. The Scat Pack later expanded to include the Challenger R/T, Challenger T/A, and Demon 340 after they were introduced. Tearing across the bottom of every print ad was a cartoon bee with racing helmet, big grippy tires, and a V-8 engine mounted on its back. The tag line was "Scat Pack... The Cars With the Bumble Bee stripes."

Like any good marketing effort, the ads weren't just directed toward obvious buyers. Much of the Scat Pack

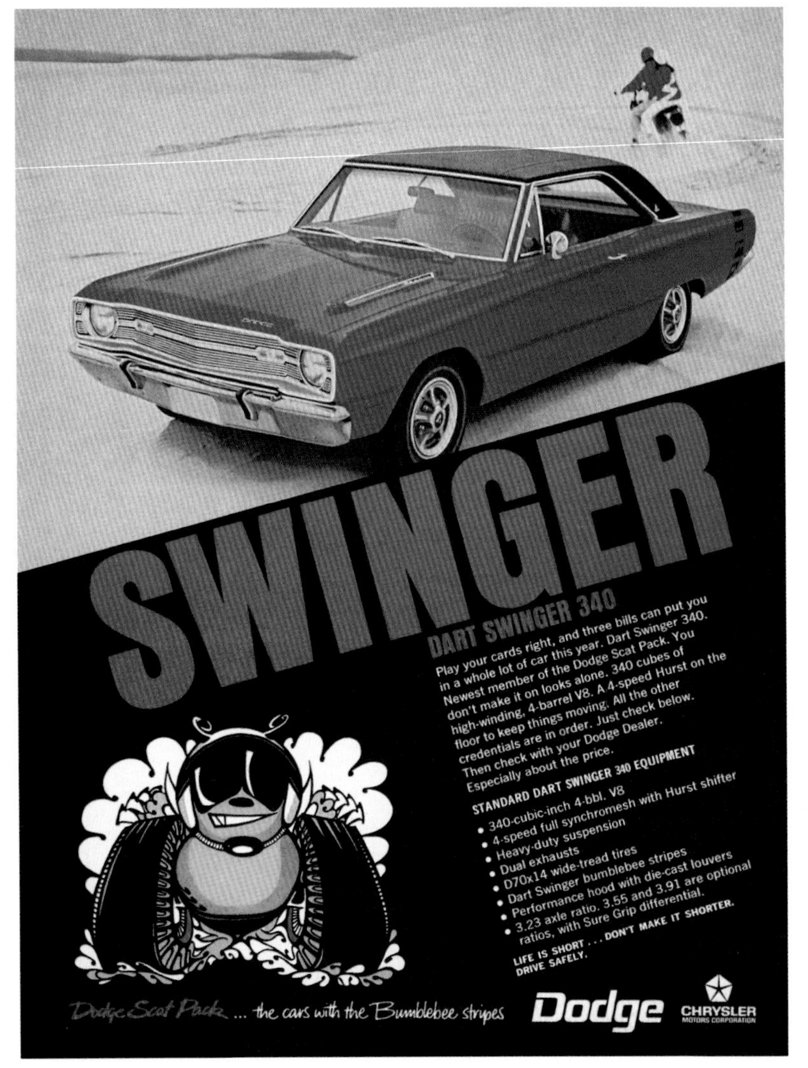

The new addition to the Dart line in 1969 was the Swinger. Although the Dart line already had the GT and GTS to cover the performance bases, the Swinger 340 was introduced as the stripped-down, lowest-cost performance car. Dodge grouped the Swinger in with its other Scat Pack musclecars for maximum advertising impact.

At Plymouth, the musclecar line was labeled the "Rapid Transit System." There's nothing subtle about clouds of tire smoke pouring from the rear wheels of a musclecar. Plymouth pitched the 1970 Duster as "the little car that could," and listed its goals for the car. "First it would have to move, really move—cut a 13/14-second quarter, pure stock."

paraphernalia was easy for kids to collect. Besides the print ads, interested parties could join a Dodge Scat Pack Club for $5.95 for year. Members received a membership card, a bumper sticker, a decal, a club patch, a tune-up tips folder, Chrysler's "Hustle Stuff" performance parts catalog, a Scat Pack accessories folder, a "pocket-pack" all-weather racing jacket, a membership card, an auto racing guide, and the monthly Dodge Performance News. The Scat Pack existed through the 1971 model year.

Linking all the performance models together worked well for Plymouth, too, starting in late 1969 as the 1970 models were rolled out. Its version of the supercar gang was called the Plymouth Rapid Transit System. "Everybody offers a car. Only Plymouth offers a system," was the promise. The "system" included the Road Runner, GTX, 'Cuda, Duster 340, and Sport Fury GT. The Plymouth advertising prominently featured such winning race car drivers as Ronnie Sox and Buddy Martin, Tom "Mongoose" McEwen, and Richard Petty. They, too, offered brochures, stickers, and posters.

One Scat Pack brochure from 1970 featured California-based Funny Car drag racer Charlie Allen, who pushed the Dart Swinger 340. "Save your cash fellas. The giant killer is here," he advised. "Dart Swinger 340 doesn't have crazy foreign names or cartoon animals plastered all over the side [Ouch! Take that, Road Runner—Ed] but that doesn't seem to slow it down much. . . . My opinion of the Swinger 340—if you want to put your dough into 'go' instead of moldings, you'll like it. I do!"

Today's collectors actively search for surviving Scat Pack and Rapid Transit System memorabilia, and reproductions have found favorable acceptance. That's surely one way to gauge a successful advertising campaign—it makes people continue buying decades after the original product has driven off into the sunset.

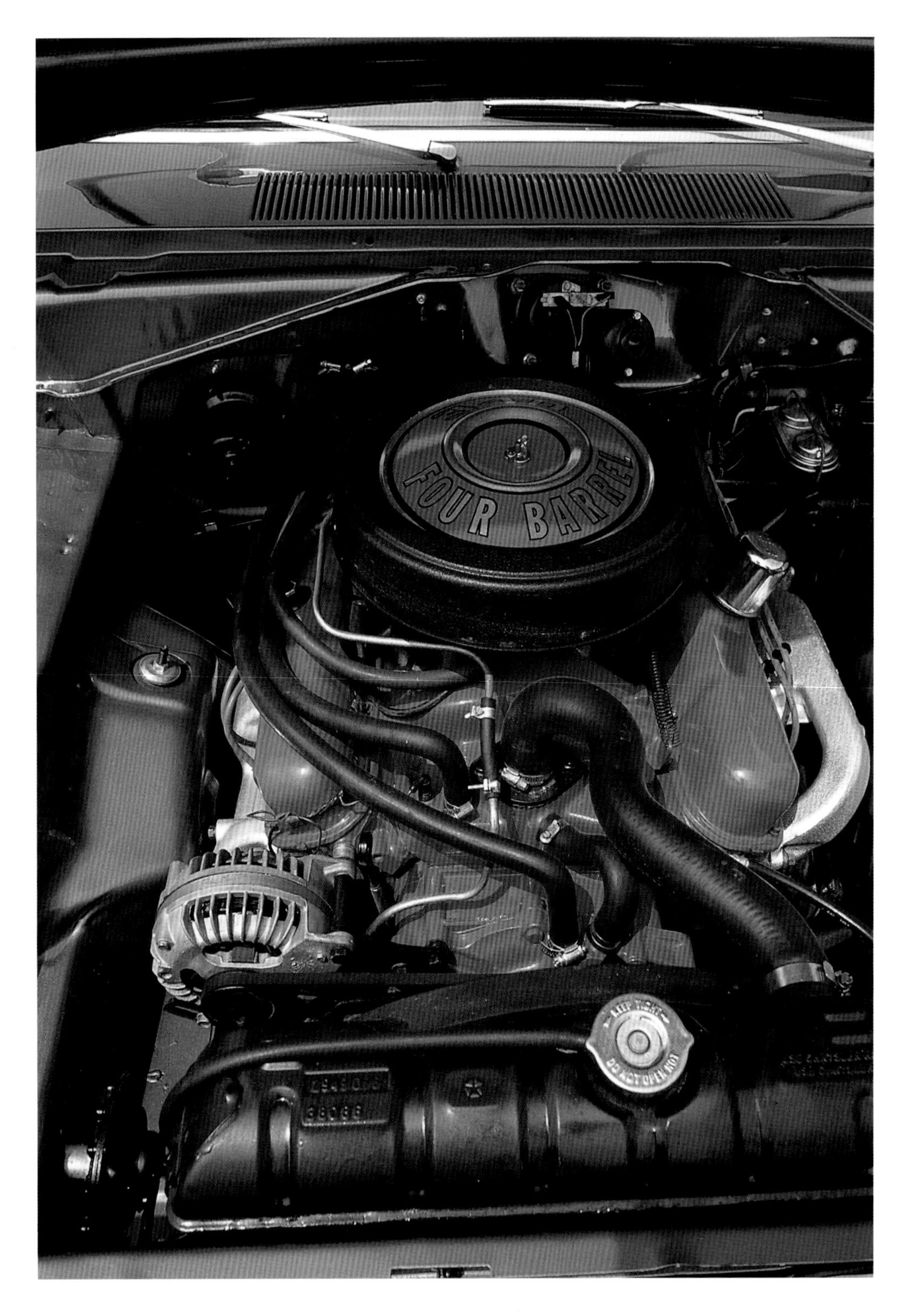

In 1969 all 340 V-8s were fitted with the same camshaft, the one that had been used solely with automatic transmissions in 1968. Horsepower remained at 275. *Photo courtesy Rob Reaser*

Continued from page 48
The Daroo II was a topless-bodied two-seater, with a shortened rear deck and lengthened hood, although the car's overall length and width were unchanged. The smooth, sloping nose recalled the Studebaker Avanti's front end. Credit for the concept went to Ron Perau's Imperial Kustoms of Tulsa, Oklahoma.

With a unique Frosted Fire color, a combination of orange and gold, and a huge air-foil-type roll bar dominating the rear of the passenger area, the Daroo II certainly played the part of 1960s show car. But the family lineage was still traceable. The Daroo II had the Dodge Scat Pack bumble bee stripe on the tail, and twin flip-up gas caps of the type seen on the redesigned 1968 Charger. There was a 340 V-8 underhood. Dodge could play with the Dart's body all it wanted, but the powertrain needed no embellishment.

Still Evolving: 1969

The Dart rolled into 1969 without major changes. The car was given the usual cosmetic makeover, with rectangular parking and side marker lamps replacing 1968's round units. Other changes included a new grille with a stronger horizontal emphasis and a revised taillamp treatment. In an apparent step backward, the safety padding that had been added to the bottom of the 1968 models' dash was eliminated. A new steering wheel design and new patterns on the vinyl seats freshened the interior.

The GT model was again the ground floor for Dart performance. The GT's trim was revised for the new year, with the most prominent feature being a new body-length molding along the beltline and relocated GT emblems. As before, the GT could be had with either the Slant Six engine, 273-ci V-8, or 318-ci V-8. The 273 V-8, a dutiful soldier since 1964, made its final appearance in the 1969 Darts.

Although a reasonably popular model since its 1963 introduction, the Dart GT was getting crowded out by musclecars with stronger engines and stronger images. Further crowding the Dart line-up were new Dart Swinger and Swinger 340 models. The Swinger 340 was Dodge's inexpensive alternative to the GTS. Produced in the mold of the low-buck Plymouth Road Runner, Swinger 340 equipment included the 340 V-8, a four-speed transmission, dual exhaust, Rallye suspension, D70 wide tread tires, plus a "Swinger" bumble bee stripe and identification and the louvered performance hood. The Swinger was only available in one body style, as a two-door hardtop. To further distinguish the Swinger 340 from the GT and GTS, the Swinger was given exclusive use of a bright red paint color for 1969. Dodge also offered a more basic Swinger that could be fitted with the Slant Six engines or the 318. Television ads boasted that the Swinger was the "Young new compact with the wild new personality."

Swinger 340 prices started at slightly over $2,800. The better-equipped GTS started at $3,226. Besides coming with a higher trim level and bucket seats, the GTS could be ordered as a convertible, or with a big-block V-8. With the Swinger 340, Dart GT, and Dart GTS in the mix, Dodge indulged in a bit of overkill with high-performance Dart models.

Ultimately, the Swinger 340's low-buck formula was the winner, with the GTS getting the ax after 1969. But the GT and GTS made a strong case for themselves in 1969, especially the GTS. The 383 V-8 was still the top engine option, and for 1969 it was upgraded to 330 horsepower thanks to a new cam. At that rating it was only 5 horsepower shy of the 383 Magnum V-8 available in larger models like the Charger.

The 340 V-8 remained the base GTS engine. For 1969 all 340s were fitted with the same camshaft, the slightly milder one used with automatics in 1968. The GTS' appearance was not substantially changed from 1968, although the bumble bee stripe was altered somewhat, with the twin stripes replaced by a fat stripe with "GT Sport" lettering. The trim on the power bulge hood was replaced by brightwork that included engine identification. With that alteration, the GTS no longer carried "383 Four-Barrel" emblems on the fenders. Dodge made slight changes to the interior trim.

On the 1969 show car circuit, the Dart provided raw material for a Custom Swinger 340. Although not the radical departure the Daroo I and Daroo II had been in 1968, the Custom Swinger got the point across. It featured a simplified grille with rectangular headlamps and center-mounted driving lights, and a blower-type "bug catcher."

The Valiant, meanwhile, soldiered on in V-100 or Signet dress. Body choices were limited to either two-door or four-door sedan. Like the Dart, the Valiant received the usual cosmetic makeover in the form of a new grille and taillamps. Underhood, the top engine option remained the 318-ci V-8.

The Valiant's strengths continued to be its relative comfort, economy, and clean, simple lines. "Last year we reported a Plymouth engineer as terming the Valiant 'the world's smallest limousine.' We thought him right then and—had he said it this year—he'd be right again," reported Bill Kilpatrick in the October 1968 issue of *Popular Mechanics*.

As the decade closed, the Chrysler A-body cars were hitting their peak. Dodge moved 197,685 Darts out of factory doors in 1969, far surpassing Ford's fading Falcon (96,016) and the dying Corvair (6,000). The larger, more powerful compacts like the Dart, Valiant, and Chevy Nova were clearly the choice of the car-buying public. And with the escalating prices and insurance difficulties of the intermediate musclecars diminishing the pool of potential buyers, the hot compacts seemingly had a clear road ahead.

DUSTER 340
GOODYEAR
POLYGLAS - E70-14

three
Dust Devils
Demons and Duster, 1970–1973

The impending release of the new E-body Barracuda and Dodge Challenger had Plymouth in an uncomfortable position. The A-body Barracuda of 1964–69 had been a nice complement to the Valiant, an inexpensive and compact performance car that was in keeping with Plymouth's image and position in the market.

The new E-body cars, however, were going to be larger and more expensive, leaving Plymouth with a gap in its product offerings. There was, after all, no Valiant-badged performance car in the line-up. People shopping for a cheap, compact muscle machine might well bypass the Plymouth dealership entirely and end up motoring away in a Chevy Nova SS or, worse from an intracorporate rivalry perspective, a Dart Swinger 340.

There was also very little Plymouth could do. The Valiant wasn't scheduled for a remake, and the division had only been budgeted a limited amount of money for a modest freshening of the Valiant's sheet metal and trim. What to do?

Throw the long ball.

Rather than spend its allocation on new trim and doodads, the Valiant team put together a design for a special semifastback Valiant, one that could compete with the hot Novas and Darts. They did this quietly, because a new model was not part of upper management's plan. Developed on a poverty budget, the new car had to carry over as much of the existing Valiant as possible, yet still be distinctive enough to justify the expense.

The solution involved stretching the car's silhouette into a mild semifastback shape. Plymouth stylist Neil Walling created the winning rear quarter-panels

The FM3 "Moulin Rouge" paint was one of the "high-impact" colors available in 1970. Just calling the cars "pink" or "green" wasn't enough—the marketing people must have had a grand old time dreaming up wild names for their paint colors, like Moulin Rouge.

The 1970 Duster was initially badged as a Valiant Duster. The Valiant designation was dropped for 1971, as the Duster had proven in its initial year that it was capable of standing on its own.

and roofline that gave the new model its personality, although, as always, styling decisions were collaborative. "In the studio there would probably be 50 to 100 proposals for a body side that would be up on the wall, that we'd all go around and vote on," explained Plymouth stylist Fred Schimmel. The final product was lower overall than the standard Valiant, and had the fat C-pillar that was reminiscent of the "ponycar" styling of the day.

On time, on budget, and on target, the sporty Valiant was given the go-ahead for production. In keeping with the cartoonish theme of Plymouth's youth-oriented models, the new car was named Duster, and its emblem was a swirling little tornado that looked suspiciously like Warner Brother's Tasmanian Devil character. The performance version was named simply Duster 340.

The Duster was only available as a two-door sedan. Its front clip utilized Valiant sheet metal, but the unique rear end treatment made the car seem like a new model from the ground up. The Duster didn't look cheap, but its low price tag drew buyers from across the demographic spectrum. Of course, attentive shoppers noticed such cost savings as flip-open rear quarter-windows, but compared to, say, a Ford Maverick, the Duster came off looking like an Imperial.

The hastily conceived design did have some drawbacks, such as the high deck lid, making loading a Duster trunk a hernia-inducing task. It was also larger than competing two-door compacts, and offered less fuel economy, although as far as Plymouth was concerned the car's greater girth was a selling point. As the Chrysler Corporation had done previously with the Dart and earlier Valiant, the Duster was pushed as a large compact. "Look out little economy cars, here comes Valiant Duster," was the advertising hook jumping from television screens.

In keeping with the Duster's mission, the interior was on the basic side, although options like bucket seats, the long center console, and the "Tuff" steering wheel provided familiar surroundings for the Mopar enthusiast. Also familiar was the Duster's instrument panel, which was lifted from the previous year's Barracuda. The Duster 340 came with the Rallye instrument cluster, which placed a small tachometer between the two main instrument binnacles.

For young musclecar buyers, the Duster 340 offered the advantage of being just another Valiant model, at least to flinty insurance agent eyes. Valiants were hardly irresponsible, teen-killing supercars,

The corporate console, long and wide enough to present a five-course meal, was an option on the Duster 340, as were bucket seats. The highly optioned car pictured here has the Interior Decor Package, Light Package, pedal dress-up, AM radio, Day/Nite mirror, color-keyed carpet, and the "Tuff" steering wheel. The Tuff wheel was released halfway through the 1970 model year and is consequently very rare on early Dusters, although it became a popular option in later years.

right? But the 340 provided just as much performance in the Duster as it did in the Dart. In a March 1970 *Hot Rod* test of the 340 Duster, Steve Kelly noted, "If the Plymouth Duster doesn't fit a logical category, such as compact or something like that, it at least qualifies as one of the best, if not the best, dollar buy in a performance car."

The magazine recorded a 15.38-second quarter-mile at 91.55 miles per hour, slower than expected for a 340-powered A-body. Their mediocre timeslip was attributed to a poorly adjusted clutch and stiff transmission, revealing both the state of Chrysler's quality control at the time, and the thrashing press cars routinely received.

The 1970 340 V-8 was not noticeably changed from the 1969 edition. Horsepower was still at 275.

The cartoonish Duster 340 sticker was a fitting complement to Plymouth's other youth-oriented, cartoon-inspired performance car, the Road Runner. Fewer than 5 percent of Duster 340 buyers in 1970 opted for the rear spoiler option.

"Criticism of some features is warranted, yet there's no getting away from the low-dollar appeal of a Duster," Kelly wrote. "It looks good, with an overall shape that attracts most passersby, rides in an acceptable fashion providing the jounce action doesn't disturb a person too much, and is a whale of a starting point for a minisized muscle car."

Other magazines had better luck coaxing representative timeslips out of the Duster 340. Both *Car and Driver* and *Car Life* scored the expected sub-15-second elapsed times. Plymouth made sure buyers knew the potential of the 340 Duster in one of the greatest print ads from the musclecar era. In a three-page mini–road test, Plymouth hired drag racer Ronnie Sox and car builder Buddy Martin, the famous Sox & Martin team, to test the entire line-up of Plymouth musclecars—marketed as the "Rapid Transit System."

The cars—a Duster 340, 'Cuda 340, 'Cuda 383, Hemi 'Cuda, 440 Six-Pack Road Runner, and 440 GTX—were raced on a road course, put to braking tests, and were each run through 10 passes at a drag strip. Few ads before or since have so brazenly bragged about performance.

The Duster used in the test was equipped with a four-speed transmission and 3.91:1 axle ratio, and weighed in at 3,265 pounds. Its average elapsed time was 14.07 seconds at 100.09 miles per hour. The Duster's quarter-mile performance was nearly a half-second quicker than that of the 'Cuda 340, and three-tenths quicker than the 383 'Cuda. The 'Cuda had the mean reputation, but it also carried 300 extra pounds compared to the Duster. Of course, no other automotive magazine was able to duplicate a 14-flat for the Duster, but then, they didn't have Ronnie Sox driving.

The Duster 340 was the headliner, but most Dusters went out the door in economy trim. The base engine was the 198-ci Slant Six, with the 225-ci six optional. The 230-horsepower, 318-ci two-barrel V-8 was the next option up the list.

In a rare bit of accurate prognostication from the automotive press, Bill Kilpatrick predicted in the October 1969 issue of *Popular Mechanics*, "If it does manage to keep the price down, the Duster could be the sales sleeper of the year." Plymouth did, and their reward was production of 217,192 Dusters for 1970. The E-body Barracuda, arriving after ponycar sales peaked in the late 1960s, sold only one-quarter as many examples. The Duster 340 also outsold the performance version of the Barracuda, the 'Cuda.

The four-door Valiant, meanwhile, was carried over almost completely unchanged from 1969. A new grille was the only noticeable new exterior feature. Underhood, the only news was the 198-ci Slant Six, which had replaced the 170-ci version. It was rated at 125 horsepower. The 145-horsepower 225-ci Slant Six and 230-horse 318-ci V-8 remained as options. With

While the Dart GTS of 1967–69 had been a nicely equipped compact musclecar, the Swinger 340 was a stripped-down money-saver in the tradition of the Dodge Super Bee and Plymouth Road Runner. A bench seat and three-speed manual transmission were two ways costs were kept in check. *Reprinted with the permission of the DaimlerChrysler Corporate Historical Collection*

the creation of the Duster, the only other available Valiant was the four-door sedan. The Signet model was dropped as well.

Dart Doings

The Dodge boys, on the other hand, did what was expected with their A-body allocation for 1970—they gave the Dart a mild facelift. The stylists sliced a wedge out of the trunk's blocky silhouette, which gave the body a forward-leaning look. The Dart got a new grille, bumpers, and hood, with taillamps integrated into the rear bumper. The dash was revamped and the number of models was cut back from nine to five. The optional bucket seats came with "high back" integral head restraints standard.

For the new year the GTS was gone, and the GT almost invisible. Although harder to find than dragon's teeth, a GT option was listed for the Dart Custom model. The 383 big-block V-8 did not make it into the 1970 Darts. The Swinger 340 was the only surviving performance Dart. It was given a couple visual tweaks, such as twin, functional hood scoops with "340" emblems on the sides. A black hood treatment was optional. Dart Swinger emblems were placed high on the quarter-panels, immediately below the rear quarter-windows.

The Swinger 340 was once again cast as a low-buck alternative to more expensive musclecars. The GTS had been a more complete package, with a smattering of creature comforts and a four-speed

340
1970
F70-14
GOODYEAR

The performance Dart received dual hood scoops for 1970, with engine call-outs on the side. A black-out hood was optional.

Inset: The 1969 Swinger 340 had relied upon a unique bumble bee stripe on the car's tail to get its message across, but in 1970 the Swinger received its own unique badging on the quarter panels.

The 1970 Dart Swinger 340, with two large hood scoops and bright color choices, cut a higher profile than the late, great Dart GTS. The famed Dodge bumble bee stripes made their final appearance in 1970.

Part of the Dart's 1970 restyle included a new rear bumper with integral taillamps, and forward-leaning sheet metal at the rear end.

transmission as standard equipment. The Swinger 340 came standard with the cheapo three-speed manual transmission, which accounted for its price decrease, and just the bare essentials otherwise. The price was right, however. "If you can find a hotter performance car for less than $2,808, buy it," ads advised. The 1969 GTS had started at $3,226.

Further down the Dart family tree, the powertrains were shared with the Valiant. The decade-old 170-ci Slant Six was retired in favor of a 125-horsepower, 198-ci version. The 225 Slant Six, at 145 horsepower, and the 230-horse 318 remained as options.

While the Dart and Valiant four-door were not radically new for 1970, they held a strong position in the economy car market. The A-bodies usually commanded a third of the compact market by themselves. Both were comfortable, reliable choices at a time when the economy market was remaking itself. Ford was starting over for 1970, with the new Maverick taking over from the Falcon. Midyear, the Falcon name was transferred to a stripped-down Torino, and then quietly put to sleep thereafter. The 1971 Pinto was scheduled to be the new decade's version of the 1960 Falcon. At AMC, the Rambler had morphed into a more contemporary Hornet, and the unusual Gremlin was released to fight for compact-leaning shoppers.

As 1971 rolled around, even Chrysler was trying new tactics. The corporation introduced the Plymouth Cricket and Dodge Colt, rebadged British Austins and Mitsubishi imports respectively. Japanese automakers, particularly Toyota and Datsun, were establishing themselves as serious players. Many of the new breed of compacts and subcompacts used four-cylinder engines and provided better fuel economy than the average six-banger American compact. The A-body's strength in the face of this changing market was a testament to the fundamental soundness of the cars.

The addition of the Duster churned up the waters considerably, though. After the Duster's huge debut year, with production of more than 217,000 cars, Dodge found itself in the unaccustomed position of A-body runner-up. Some Dodge dealers itched for their own version of the popular compact, especially since the Challenger was not meeting sales expectations.

"The Duster had proven very successful, just taking a very boxy-looking Valiant, the way that it was able to make a more flowing shape out of what had been a fairly boxy design, seemed to be well-liked," recalled Plymouth stylist Fred Schimmel. "The public was buying Dusters quite well."

Although many within Plymouth resisted, Dodge won the argument and was hastily penciled in to get its own "Duster." Dodge's version, the Demon, was introduced for the 1971 year model. A Duster clone, the car was given just enough Dodge character to separate it from its Plymouth cousin. The Demon kept the Duster roof and semifastback body, but used a Dart grille and front-end sheet metal. Of course, by hastily joining the Dart front sheet metal with the Duster rear, Dodge created a mismatch between the shape of the front and rear wheel openings. Such shortcuts led to Chrysler's reputation for shabby quality control in the 1970s.

Changes to the 1971 Duster 340 were certainly eye catching. The car received its soon-to-be-famous body-length 340 side stripes and a new segmented grille. *Reprinted with the permission of the DaimlerChrysler Corporate Historical Collection*

DEMON 340
...the performance is a lot more than painted on.

Some people today seem to be building performance cars for your kid sister. They offer a brand-new stripe, but the same old Six.

Dodge kids you not. This may be our lowest priced performance car, but you'd never know it from the way it keeps up with the big boys.

This year, you get our high-revving small V8. With new frenched rear lights and a clean-looking grille. All this runs on heavy-duty torsion-bar-sprung Rallye Suspension; heavy-duty shocks; big brakes; and a slick, full-synchro floor-mounted box. Add a readable speedometer with resetable trip indicator and a sanitary, roomy, all-vinyl interior, and you're ready to roll.

Demon 340. Compared to some of the other new ones you've seen, it's a nice honest car . . . with a nice honest price.

STANDARD EQUIPMENT

340-cu.-in. 4-bbl. V8 (premium fuel) □ Vinyl front bench seat—Blue, Tan or Black □ Body side tape stripe—Black or White only □ Ventless door glass □ Custom-contoured outside door handles □ 2-speed windshield wipers □ Dome lamp □ 150-mph speedometer, with trip odometer □ Fuel, alternator, temperature, and oil pressure gauge □ Cigarette lighter □ Heater/windshield defroster with 2-speed fan □ Transistorized regulator □ 3-speed manual transmission, fully synchronized with floor-mounted shift lever □ Rallye Suspension Package (includes heavy-duty torsion bars, heavy-duty rear springs and sway bar with heavy-duty shock absorbers) □ Brakes: 10" x 2¼", front; 10" x 1¾", rear □ E70 x 14 wide-tread, bias-belted tires □ 14 x 5.5J wheels □ Dual exhausts □ 17-gallon fuel tank.

Although Demons were economy cars at heart, Chrysler offered enough performance options to turn the 340 models into competitive musclecars. Optional dual hood scoops, hood pins, and rear deck spoiler shouted the car's intentions in no subtle terms.

Deriving from the Duster, the Demon rode on the Valiant's shorter 108-inch wheelbase, rather than the Dart's 111-inch wheelbase. The Demon had unique taillamps, plus its own cute cartoon mascot (one that Dodge planners would later wish they had given more thought).

Although stylish looking, in its way, the Demon—like the Duster—was designed as inexpensive transportation. It used cheap, pivot-type rear quarter-windows. The base engine was the 198-ci Slant Six, with three-speed manual transmission. "The Demon will be a price leader—a value package with particular appeal for younger families, college students and women who work," said Robert McCurry, Dodge's general manager at the time.

Dodge also created a performance version of its new car, called simply Demon 340. Like the other A-body muscle editions its standard equipment included the 340 V-8, dual exhaust, a three-speed manual transmission, a Rallye instrument cluster with 150-mile per hour speedometer, Rallye suspension, and E70 x 14 tires. Visuals included a side

In 1970 and 1971, Plymouth assembled the Rapid Transit System Caravan of muscular show cars, of which this unique Duster 340 was a part. Other cars in the Caravan included a 1970 Road Runner, followed by a restyled 1971 Road Runner, a 440 'Cuda, and a replica of Don Prudhomme's Funny Car. The cars toured the country constantly, stopping at car shows and dealerships, and were supplemented by gee-whiz goodies like cut-away engines. *Reprinted with the permission of the DaimlerChrysler Corporate Historical Collection*

The 1971 Duster cast aside its Valiant nameplates and low-key look, offering a lot more visual impact to go with the 340 V-8. This black-out hood treatment with "340 Wedge" identification was one option. The Wedge designation referred to the wedge shape of the V-8's combustion chambers.

stripe in black or white, with a black-out hood, dual hood scoops, and a rear spoiler optional.

The Demon 340 was quickly included in Dodge's "Scat Pack" marketing campaign. In a Scat Pack ad that year, Dodge pushed the Demon 340 by thumbing its nose at the newer "decal musclecar" competition. "Some people today seem to be building performance cars for your kid sister," they judged. "They offer a brand-new stripe, but offer the same old Six."

That sales tactic was fairly brazen, considering Dodge was at the same time offering a decal musclecar of its own, complete with "same old Six." By 1971 even compact-sized musclecars with small-block engines were proving difficult for young adults to purchase, thanks to insurance surcharges tacked onto any car that offered even a whiff of high performance. The Big Three's strategy for dealing with this buyer's *continued on page 70*

With new body-length side stripes and hard-to-miss 340 identification on the quarter panel, the 1971 Duster 340 traded in its conservative Valiant duds for a rock 'n roll image. Vinyl tops were a popular option.

Buyers of the 1971 Duster 340 could choose between a split bench seat with fold-down armrest or optional console with floor shifter.

Inset: Although still rated at 275 horsepower, the 1971 340-ci V-8 was tweaked a bit to improve emissions. The Carter AVS carburetor of previous years was replaced by a Carter ThermoQuad with smaller primary bores for a leaner fuel mixture during steady-throttle cruising. The ThermoQuad offered good performance and economy, but more restrictive exhaust manifolds for 1971 negated any potential performance gains.

The color-keyed striping extended to the rear panel in 1971.

The Rallye Instrument cluster included a 150-mile per hour speedometer—hopelessly optimistic for a Duster, but a fun feature nonetheless.

Continued from page 66
dilemma was to offer tape-striped cars that looked like big, bad musclecars, but were a lot more tame underhood. Ford's entries were the Maverick Grabber and Mercury Comet GT, while Chevy jumped in with a Heavy Chevy Chevelle and Rally Nova. Pontiac's Le Mans T-37 could be ordered with hefty engines, just like the GTO, but the base engine was a 350 V-8.

Efficient, if on the small side, the 1971 Rallye Gauge cluster included a fuel and temperature gauge, oil pressure gauge, and ammeter all in one binnacle. A tachometer was mounted in the center of the dash between the gauge cluster and the 150-mile per hour speedometer.

Dodge's effort was called the Demon Sizzler. Like the others in its class, it was hard not to notice. Standard equipment included Rallye wheels, a color-keyed grille and outside "racing" mirrors, loud Sizzler stripes, and a black hood treatment. The Sizzler package also included the Tuff steering wheel, carpeting, plaid cloth and vinyl bench seats, bright drip rail and wheel lip moldings, and 6.95 x 14 whitewall tires. The standard color was silver, but in keeping with the "look at me" image of the car, paint choices included Chrysler's wildest colors, such as Top Banana, Citron Yella, Plum Crazy, and Hemi Orange.

Plymouth had a companion model to the Sizzler, named, in a fit of redundancy, the Duster Twister. Both could be upgraded with a cassette stereo tape player and recorder that was added to the option list in 1971.

What kept the cars affordable was what was underhood—or more accurately, what was not underhood. The Sizzler and Twister were not available with the 340 V-8. The cars could be ordered with the 198-ci Slant Six, the 225 Slant Six, or the 318 V-8.

The Duster 340, meanwhile, cut a much higher profile in 1971. The low-key sleeper look was jettisoned. The 1971 Duster 340 featured large billboard graphics along its flanks, and a new grille. An optional black hood treatment had a huge "340 Wedge" graphic affixed to the right side of the hood. Dusters could even be ordered with a fabric sunroof. The car's name was shortened from Valiant Duster to just Duster. In one year the Duster name proved strong enough to stand on its own.

The 340 V-8 was still rated at 275 horsepower, although the 1971 exhaust manifolds were more restrictive than earlier editions. The potent small-block was also fitted with a Carter ThermoQuad carburetor, which was more easily tuned for emission controls.

Primitive emission-control devices were introduced more widely in the 1971 models. The evaporative control system used previously only on the California models became standard on all 49-state cars. California cars meanwhile were fitted with a solenoid valve in the vacuum line to the distributor to reduce spark advance, thereby decreasing nitrous oxide (NOx) levels.

Bold stripes notwithstanding, the new grille was arguably the most striking feature of the 1971 Duster 340. The sharply segmented grille gave the Duster 340 a distinctive face.

Halfway through the 1971 model year, Plymouth introduced a Duster for those who wanted musclecar looks on a six-cylinder budget. The new Duster Twister got the look right with a Duster 340 grille, racing mirrors, side stripes, Rallye wheels, a black-out hood, and hood scoops lifted from the Dart Swinger 340. The largest available engine was the 318-ci V-8.

The use of the Warner Brothers' Road Runner cartoon for Plymouth's low-buck musclecar proved successful enough that the idea spread to other performance cars in the fleet. The Duster used a swirling dust devil cartoon to establish its identity, and likewise Dodge's Demon was given a cute but mischievous character of its own. Not everyone was amused, however, and the Demon cartoon drew protests from religious leaders. After only two years, the Demon name was dropped in favor of Dart Sport.

With Dodge encroaching on the Duster's territory, Plymouth was given a little payback in the form of a two-door hardtop, similar in mission to the Dart Swinger. In fact, it was a Dart Swinger with a Valiant nose. Named the Plymouth Scamp, it rode on the 111-inch wheelbase of the Dart. After being cut to the bone in 1969, the 1971 Valiant line consisted of the 108-inch wheelbase four-door sedan, the 111-inch wheelbase Scamp, and the Duster.

With a cute name like Scamp, television ads pitched the new Plymouth to women. "That's why this lady drives a Scamp," was the theme song, sung to the tune of "The Lady is a Tramp." Like all the A-bodies, the Scamp was also positioned as a low-cost car for first-time buyers and people on a limited budget.

The 1972 Fade

In 1972 the effects of unleaded gasoline and federal emission controls made a genuine impact on performance. Where once a PCV valve and a tune-up

Although a formidable performance car when equipped with the 340 V-8, most Demons were purchased with economy car duty in mind. The base package included Slant Six engine, three-speed manual transmission, and dog-dish hubcaps.

Gauging the Competition—1970

In 1960, the Valiant's debut year, the new little Mopar's competition included the Falcon, Corvair, Studebaker Lark, and AMC Rambler. By 1970, the Corvair and Lark were history, the Rambler had become a Hornet, and the Falcon name was in the process of being attached to a stripped-down Torino. Not that Ford, General Motors, and AMC had given up on the compact market—for the new decade there were many choices, although thoroughly conventional ones. The Big Three had been much more daring a decade earlier.

But each new car brought something to the table. The AMC Hornet provided cut-above interior appointments. The AMC Gremlin was, well, odd. The Ford Maverick was a penny-pinching price leader. The Chevy Nova most closely matched the Duster in terms of engine availability and all-around usefulness. Of course, in 1971 the economy car market was reconfigured by the arrival of the American subcompacts like the Ford Pinto and Chevy Vega, but the traditional compacts didn't suffer from the competition.

Figures for 1970 models:	**Nova**	**Hornet**	**Maverick**	**Duster**
Wheelbase (in.)	111	108	103	108
Overall length (in.)	190	180	179.4	189
Curb weight (lbs.)	2,919	2,705	2,411	2,790
Base engine (ci)/cyl.	153/4	199/6	170/6	198/6
Top engine option (ci)/cyl.	402/8	304/8	200/6	340/8
Max. horsepower	375	210	120	275
Two-door base price	$2,176	$1,994	$1,995	$2,172

On the performance front, only the Duster 340, Nova SS, and Dart Swinger 340 were players in the compact segment. Ford didn't even offer a V-8 in the Maverick, initially. The first Hornets could be ordered with AMC's 304-ci V-8, but with two-barrel carburetor and 210 horsepower, they were hardly the stuff of dreams. The only other relatively small-size American cars buyers could choose from were the ponycars, like Ford's Mustang, the Chevy Camaro, Pontiac Firebird, AMC Javelin, or Chrysler's own Challenger and Barracuda. All fine choices, but the ponycars required owners to sacrifice rear seat utility, not to mention shelling out more bucks. And, as many a buyer discovered, price was often what made the Duster 340 and Swinger 340 the winning choice.

Musclecar Base Prices, 1970
(All prices factory A.D.P—Manufacturer's suggested advertised delivered price. Does not include shipping charges.)

Make/Model	Factory A.D.P. Price
Plymouth Duster 340	$2,547
Dodge Dart Swinger 340	2,631
Chevrolet Nova w/ SS option	2,794
Plymouth Road Runner	2,896
Dodge Coronet Super Bee	3,012
Buick Skylark GS350	3,098
Chevrolet Camaro w/ SS option	3,129
Plymouth 'Cuda	3,164
Chevrolet Malibu w/ SS396 option	3,254
Dodge Challenger R/T	3,266
Pontiac GTO	3,267
Ford Torino Cobra	3,270
Ford Mustang Mach 1	3,271
Buick Skylark GS455	3,283
Pontiac Firebird Formula 400	3,370
AMC AMX	3,395
Chevrolet Camaro w/ Z28 option	3,412
Plymouth GTX	3,535
Dodge Coronet R/T	3,569
Dodge Charger R/T	3,711
Ford Boss 302 Mustang	3,720
Mercury Cyclone Spoiler	3,759
AMC "Mark Donohue" Javelin	3,995
Plymouth Road Runner Superbird	4,298
Pontiac Firebird Trans Am	4,305
Oldsmobile Cutlass 442	4,467

The 1972 Duster 340 picked up a small tornado cartoon at the rear of the body side stripe, but otherwise it is hard to tell a 1972 from a 1971 at a glance. Under the hood the 340 lost compression, and was down-rated to 240 horsepower. *Reprinted with the permission of the DaimlerChrysler Corporate Historical Collection*

comprised the whole of Detroit emission control, the new rules had teeth and required substantial changes. The 340-ci V-8 was noticeably detuned for 1972, with the compression ratio lowered from 10.5:1 to 8.5:1. The 340's performance-geared cylinder heads were replaced by the 360-ci two-barrel V-8's heads, which contained smaller valves. The changes choked horsepower out of the engine, plus Detroit automakers switched from a gross horsepower rating to a more realistic net figure. The 340 was downgraded to 240 horsepower.

Not all the news was bad, however. The 340s were fitted with electronic ignition systems for the new year, which helped with emissions without hurting performance, not to mention making tune-ups easier.

Plymouth also tried to expand the Duster's appeal by offering something for everyone. The new addition

After the Demon name was retired at the end of the 1972 model year, Dodge's Duster clone was renamed the Dart Sport. The 1973 models were also given a facelift that included a new hood, grille, and front bumper.

Large, blocky bumper guards helped the 1973 Dart line meet new federal 5-mile per hour bumper standards. The taillamp design was new for 1973 as well.

for 1972 was the Gold Duster. The Gold Duster option included such posh features as deluxe wheel covers with whitewalls and an exclusive gold color, plus Gold Duster identification and a reptile-grain vinyl top. The option could be teamed with either the 225-ci Slant Six or the 318 V-8, but not the base 198 Six or 340 V-8.

Besides the diluted performance, Mopars of the early and mid-1970s suffered in the quality control department. Labor unrest was one cause, with often caustic relations between workers and management leading to less-than-conscientious assembly of the product. Nineteen seventy-two was an especially contentious year for automakers, with strikes at GM shutting down production for months.

As Jim McCraw found in his June 1972 *Super Stock & Drag Illustrated* review of a 1972 Demon with 340 and sunroof, the car had quality control problems everywhere—loose seats, ill-fitting dash inserts, poor window seals, etc. Plus, the 340 V-8 was so loud it deserved special commentary. "When you consider some of the early 1950s British sports cars and how noisy, leaky, uncomfortable, harsh they were and how popular they were with enthusiasts of the time, you begin to think that perhaps the Demon 340 customer views his machine as a real 'macho' piece, full of mechanical flavor," he wrote.

"The manually operated sunroof on the Demon was a good deal all around. The latch was a bit stiff, but once we had flipped it around 180 degrees to unlock the sliding mechanism, it moved very easily and could be locked in any position along its travel from just a couple of inches to full opening, which is considerable. With the sunroof all the way back, there's a 28x28-inch hole in the roof that extends over the back seat, admits plenty of sunlight, and not much wind.

With the retirement of the Demon name, the fastback Darts were rechristened Dart Sports.

The "Tuff" steering wheel was an option that almost single-handedly elevated Duster interiors from cheap to inviting. A Hurst shifter was still a Duster 340 feature.

The 340 V-8 made its final appearance in the 1973 models. Although the previous year's drop to an 8.5:1 compression ratio, along with further leaning-out of the fuel mixture, had dropped horsepower from 275 to 240, the 340 V-8 was still one of the top performance engines of its day. It was replaced by a 360-ci version of the small-block in 1974.

The fit and finish of the sunroof inside and out were very good, which probably means that it was installed by an outside firm after the car was built."

Being a drag racing magazine, they naturally took the car to the track, where it clicked off times ranging from 15.49 to 15.18. After removing the air cleaner lid, opening up the hood scoops, and playing with tire pressures, they recorded times in the 14.90 range—not bad for a low-compression V-8, especially compared to the Demon's contemporaries. But the times were still several tenths off from what Dart GTS 340s had been running just three years earlier.

Looking Up and Down: 1973

Performance may have been falling, but the A-body cars made up for that in other ways. Big improvements for the A-bodies arrived in 1973. The Dart and Duster received new front suspension systems, with new upper and lower control arms and a heavier-duty upper ball joint. A larger diameter spindle on disc brake cars allowed for a 4 1/2-inch diameter bolt circle, which improved the choices of aftermarket wheels available for the cars.

A revised steering system allowed parts sharing of the steering gear, pitman arm, and sector shaft between the A-bodies and the E-body Challenger and Barracuda. The V-8 cars were given front disc brakes standard. The rest of the Chrysler engines were fitted with electronic ignitions; the 340 had received the new system in 1972.

The changes were not just out of sight under the skin, however. The A-bodies were given new grilles, taillamps, fenders, and hoods—a necessary restyling to accommodate the large federally mandated heavy-duty bumpers. Formerly optional, bumper guards were made standard equipment.

At least one change for 1973 had nothing to do with engineering or federal mandates, and everything to do with public relations. Naming Dodge's Duster clone the Demon had always been a cheeky proposition. It was all in fun, with a cute little cartoon Demon to emphasize that point. But inevitably, someone was bound to get offended by naming a mainstream Detroit auto after a denizen of the underworld. Some people take that sort of thing seriously, and in the case of the Demon, they did.

Likewise, corporations are sensitive to bad press and unfavorable public opinion. As letters trickled in and a few vocal opponents found more mainstream forums for their views, Dodge decided to avoid the whole controversy. It dropped the Demon name and imagery, and renamed the car the Dart Sport. Nobody objects to darts, do they? Another factor was that the Duster had sold in huge numbers, while the Demon had posted far fewer sales. In 1971, for example, Plymouth built more than twice as many Dusters as Dodge did Demons. Perhaps the Dart nameplate was worth more than originally assumed.

The renamed and remolded Dart Sport was examined in the February 1973 issue of *Motor Trend* in a comparison test against the Chevy Nova hatchback and AMC Hornet Hatchback. The Dart Sport was praised for its instrument layout and handling, although the *Motor Trend* editors clearly found room for improvement. "The basic flaw in the Dart is that the good things are done so well, the shortcomings are amplified," writer Jim Brokaw noted. "The high-back bucket seats, for example, are well-padded, of good quality vinyl, and reasonably comfortable. But the driver sits very low in the car, and the seat angle gives the impression that you are about to slide off the front. You never do, of course, but you worry about it.

"The low seating position combines with a rather steep angle on the steering wheel to produce an uncomfortably high wheel. I'll bet the wheel angle fits the bench seat a lot closer."

With the success of the Duster, and to a lesser extent the Demon, Chrysler Corporation continued to try to make the cars appealing to as many groups of people as possible. One new addition for 1973 was the Space Duster. Although moon landings were still fresh news, the Space Duster was not named for any outer space achievements. It earned its title because the trunk wall and rear seatback folded down, offering a 6 1/2-foot flat surface from the rear of the trunk to the back of the front seats. The feature was, naturally, introduced

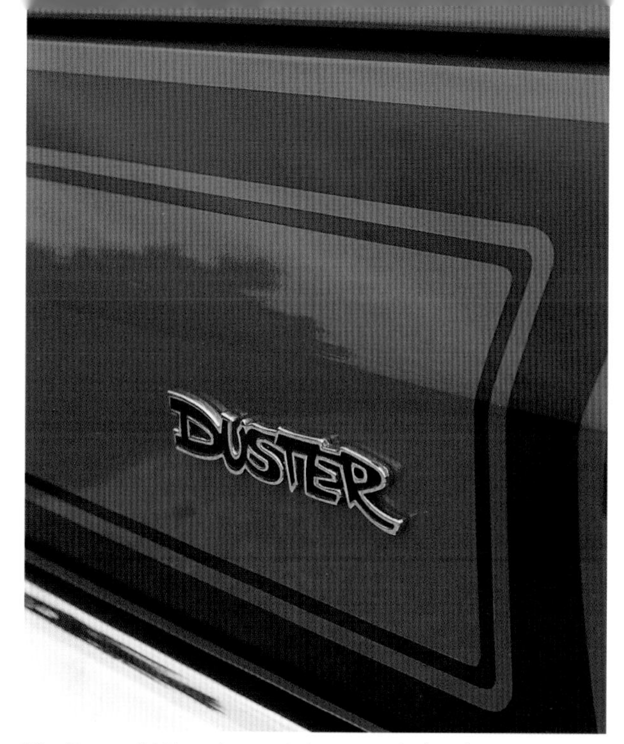

The Duster 340's unique striping extended to the rear panel.

on the Dart Sport as well, to great success. More than a third of Dart Sport buyers opted for the feature, crowed Dodge general manager Robert D. Loomis in a press release. The folding seat provided the benefit of a hatchback's cargo space, without leaving all your valuables in plain view of any passing thief. The "convertriple" package teamed the fold-down seat with the manually operated sunroof.

The Duster Twister and Gold Duster packages continued for 1973. A lesser-known new package was the Dart Sport Rallye. Introduced at the Chicago Auto Show, and phased in late in the model year, the Rallye package included, not surprisingly, the Rallye suspension and Rallye wheels, E70x14 white letter tires, disc brakes, power steering, the Tuff steering wheel, 318 V-8, and four-speed transmission. "Dodge" lettering on the quarter-panel gave the car a unique graphic signature.

As Loomis explained it: "In an industrywide shift of many buyers away from the muscle or specialty compact car, there has been a trend toward the smaller, less powerful car which is both economical and fun to drive," he said. "And the Dart Rallye is just such a car for men and women who enjoy driving."

Rather than the musclecar enthusiast, Dodge was chasing after sports sedan buyers. "There always has been and still is a large number of aficionados, whose greatest enjoyment from their car is in the very pleasure of driving," Loomis said at the time. "These are the

Although the Duster was an economy car at heart, by 1973 buyers could load up on luxuries like a rear window defogger and a map light. Toggle switches were mounted along the bottom of the dash.

340

people who appreciate the engineering that goes into smoothly meshing gears. They get satisfaction from grasping and operating a stick shift manual transmission, from the feel of smooth cornering, snug braking, and from a car's balanced lean on a winding road."

An advertisement for the 1974 Dart Sport Rallye expanded on the car's focus. "Dart Sport Rallye wasn't made for those who buy on cubes alone," the ad explained. "A super car with a super price, it is not. But boring, dull or commonplace, it isn't either. The power-to-weight ratio works out to a shade over 20 pounds per horsepower . . . and that puts it in the top 10 percent of the class."

The Rallye package hardly turned the Dart Sport into a BMW 2002, but the emphasis away from tire-smoking performance was understandable. By 1973 sales of musclecars had dwindled down to a fraction of their former numbers. The Barracuda and Challenger had been on a long, slow slide, and the Charger Rallye never was as popular as the old R/T version. The Super Bee was two years dead. In 1972 Plymouth sold fewer than 10 percent the number of Road Runners it had produced in the car's peak year of 1969.

Other manufacturers faced the same problem. The Firebird was nearly canceled after a dismal 1972 sales year. Only a paltry 5,807 buyers opted for the once-invincible GTO that same year, and even fewer in 1973. Ford's Mustang sales had steadily dropped, and AMC's Javelin was close to the chopping block.

Meanwhile, sportier touring-type cars like the Datsun 240Z and 510, aforementioned BMW, and Volvo 140-series had posted steady gains. Ponycar pretenders like the Toyota Celica were also tempting customers who might once have purchased a Camaro or Barracuda.

Expanding the available offerings to include Gold Dusters and Space Dusters, Dart Sport Rallyes and Duster Twisters, Dart Swingers and Plymouth Scamps, let the A-body be all things to all people. A musclecar version was still around, but so were sporty-image packages, economical versions, and family sedans. This A-body diversification strategy was one reason Duster and Dart sales stayed strong at a time when so many other American cars were losing ground. It helped that the Valiant and Dart already had a reputation as economical transportation during a time of rising gasoline prices. Similar cars from other manufacturers, such as the Chevy Nova and Ford Maverick, also survived the down times in good shape.

Like the Dart Sport, the 1973 Dusters were given new, chunkier nose and tail treatments. The deck spoiler on this car is an owner addition from a 1972 model.

Demon 340
POLYGLAS - E70-14
GOODYEAR

four

Haunted by Demons

Mr. Norm's Hot Rods

When people fondly recall the glory days of their musclecar youth, they aren't just reminiscing about the machinery. They remember the people, the places, the smell of tire smoke and burning clutches, and the scary thrill of oversize V-8s propelling barely controlled, insanely powerful cars toward triple-digit speeds. They remember the scene that surrounded packs of hot rods on weekend nights.

They remember guys like Norm Kraus.

If you were a Dodge Demon guy, you knew about "Mr. Norm," as he was known in his advertising. For that matter, if you were a Dart, Challenger, or Super Bee guy you probably knew about Mr. Norm's Grand Spaulding Dodge in Chicago. While most car dealers gave young high-performance enthusiasts the same sort of treatment they'd give a wet dog wandering into the showroom, Norm Kraus embraced them. He loved the high-performance scene, reveled in it. Had fun with it. "With us, kids were kings," Kraus said. "All our performance salesmen were from 21 to 27."

It was, of course, all about selling cars. All the fun, the hot cars, the hours at the drag strip, meant nothing if it didn't translate into sales. Grand Spaulding's formula for success was to cater to the performance market while building a reputation for selling cars that ran just a bit better than those of the typical Dodge dealer. Kraus and crew made it all work, to the point where during a five-year span in the late 1960s and early 1970s, Grand Spaulding Dodge moved from being the Number One dealer of high-performance Dodges to being the highest volume Dodge dealer in the world. He eventually became the most profitable dealer in the United States, a feat that was achieved through high volume. Much of this success was achieved before Norm Kraus' 30th birthday.

Just as Grand Spaulding Dodge had given the Dart GTS an improved image, the dealership also turned the wick up on the Demon. For 1971 (shown), the answer was Six-Pack induction on the 340 V-8. For 1972, Mr. Norm bolted Paxton superchargers to his GSS Demons.

Norm Kraus (Mr. Norm) traded on his image as the "high-performance car king" to sell his highly tuned Dodges. Although he sold and dyno-tuned any of Dodge's musclecars, his dealership specialized in his unique GSS Darts and Demons.

Grand Spaulding Dodge wasn't completely alone. Other "super dealers" of the 1960s hit upon the idea of offering their own versions of the factory supercar. These dealers took the stock muscle machines and offered their own tweaks, sometimes even swapping engines to get the maximum engine in the smallest car. One of the highest profile performance dealers was Yenko Chevrolet of Canonsburg, Pennsylvania. Don Yenko parlayed his close relationship with the factory into the creation of a series of 427-powered Camaros and Chevelles. Nickey Chevrolet in Chicago was another of the power dealers that fielded its own 427 Camaros. Royal Pontiac in Royal Oak, Michigan, was famous for its Royal Bobcat conversions of the GTO, and profited greatly from the performance work. On the Blue Oval side, Tasca Ford in East Providence, Rhode Island, worked similar magic with Mustangs and Torinos. Tasca was instrumental in the creation of the 428 Cobra Jet Mustang. Grand Spaulding became best known for its run of GSS Darts and Demons.

Mr. Norm's hot car enterprise started as a used car lot in Chicago. In the late 1950s, almost by accident, he discovered that performance cars—in fact, just about anything with a V-8 and manual transmission—pulled in lots of interested young shoppers. Kraus, and his brother, Leonard, had been approached by the local Dodge representative about becoming a Dodge dealer, but they resisted initially, not keen on the Dodge product line. That initial wariness was overcome in 1962, after seeing the type of performance cars Dodge had in the pipeline. The Max Wedge V-8s had made a sale.

In 1963 the new Grand Spaulding Dodge started showing its name at the drag strip. They became a lot more involved after Kraus met race driver Gary Dyer at the track one day. The two hit it off, and ended up racing together for 10 years. Dyer signed up at Grand Spaulding as director of performance, a job that included pedaling a series of supercharged Funny Cars down the track in match races.

The dealership also started the Mr. Norm's Sport Club, of which you were automatically a member if you bought a car. The Sport Club offered a discount on parts and service, a subscription to *Drag News* magazine, a window decal for the sport club and window decal showing the car was power-tuned, a T-shirt, a trunk decal, a membership card, a list of events and contests, and bonuses for bringing in new members. A member who brought someone in to buy a car got $75. And Mr. Norm's got a new member.

"We took care of them," Kraus said of his customers. "We repaid them. One night when we raced at Rockford, we had the Buckinghams out there, and we had the beach rented, which was about a mile

Grand Spaulding Dodge utilized high-performance engine parts developed for the 1970 Dodge T/A Challenger and Plymouth AAR 'Cuda to create its 1971 Demon GSS models. The Edelbrock Six-Pack manifold and trio of Holley two-barrel carbs were available through Chrysler. Grand Spaulding rejetted the carbs, traded the vacuum-operated linkage for progressive mechanical linkage, and treated the cylinder heads to custom modifications. Mr. Norm's performance-minded dealership also swapped in a Crower camshaft and lifters, a Milodon high-volume oil pump, a Holley high-volume fuel pump, and special aluminum air cleaner.

from the track. We had about 800 members there that night. All the members, in between races, were dancing behind the stands."

Grand Spaulding also used its sales muscle to get the best deals on accessories, and ran specials on parts often. "They could buy things at our place that they could not come close to (on price) on the street," Kraus said. "I would make special deals with Keystone wheels or Cragar wheels. We used to order 100, 200 wheels at a time and get special prices. I used to pass the savings on to the customers."

Although common now, clothing was not usually found at car dealerships in the 1960s—unless you were shopping at Grand Spaulding. "We had a boutique in 1965," Kraus said. "We had shirts, ties, jackets, and in the winter time we even had gloves and boots. I took that on a dare. One of our customers said 'I bet you think you can sell clothes in here, too!'

"I said 'I bet I can.'

"He said 'I'll bet you $100.'

"I went out the next day and bought a couple display cases, and I called up and ordered shirts and ties, brought all the things in and started selling them like crazy."

Clinching Sales

Besides the race cars, club activities, and clothes, Grand Spaulding Dodge tried to build a reputation for dealing straight with its customers, and applying a personal touch.

"We sold every performance car for $200 over invoice," Kraus said. "What would happen if you

Grand Spaulding Dodge was very effective at earning Mopar enthusiasts' loyalty through such tactics as the creation of "Mr. Norm's Sport Club." Besides creating a sense of camaraderie among his customers, the club also offered discounts on parts, special events, and special decals. The Sport Club decal was accompanied by a "Dyno-Tuned" decal. A Mr. Norm's window sticker was also an ego-boosting warning to drivers of other makes.

go to the track and you've got 20 of your customers sitting, talking together, and they've all got the same cars, and they all paid a different price?"

Of course, extras like mag wheels and such added to the bottom line, and Grand Spaulding missed few opportunities to make those sales. "When a person would buy from us, they could buy a completely stock car from us. That's fine," Kraus said. "They could buy wheels and tires. They could buy with a high-performance oil pan on it. They could buy it with a cam, they could buy it with a valve job, with a distributor. They could buy it with a quick shift transmission. With a traction bar on it. With a gear. With wide-tread tires, wheels. Whatever was made for a car, we had it in the shop. That's where we got the reputation."

Mr. Norm also made an effort to talk one-on-one with musclecar buyers. "Any person who bought a car and wanted performance, I personally would sit down with that person," Kraus said. "My first question was 'How are you going to drive it? Tell me how you're going to drive it, and I will suggest what you should do.'"

Mr. Norm had other methods for making the sale, including inspiring a little awe among young customers. "We had to capture their imagination in the service department," Kraus said. To that end, the shop was strewn with black-and-white checkered racing flags, the walls covered in pictures of all the race cars Grand Spaulding had sold.

Perhaps most important were the two chassis dynos. Rolling newly purchased Dodge R/T musclecars onto a chassis dyno for power tuning was an experience the average Dodge dealer couldn't provide. Along with the distributor machines and other tuning equipment, it was quite a sight.

A chassis dyno, since it measures horsepower at the rear wheels, will naturally show a lower horsepower figure than the factory number, which is measured with an unencumbered engine on a dynamometer. This was useful for a variety of reasons.

"We'd tell the insurance company that it was really no big deal, this is not a race car, it's a stock engine from the factory," Kraus said. "We would say that it's the same engine that all the secretary of state police cars have in it. It's a cruising engine."

As Kraus remembers it, most 440-ci engines, while rated at 375 gross horsepower from the factory, put out about 190 horsepower at the rear wheels. "I'd tell them, 'Come here, I'll show you on the dyno. It only puts out 190 horsepower,'" he said. After a session of power tuning, which was performed with the customer watching, the 440 could leave the dealership with as much as 325 horsepower measured at the rear wheels.

"You had to do those things when you were in the performance field," Kraus said. "It wasn't like we were pulling a fast one. We didn't detune the car. That's the way they came in from the factory."

With such methods it was no wonder Grand Spaulding Dodge sold so many Dart GSS 383s, and GSS 440s (covered in greater detail in chapters 2 and 6). Although famous for tweaking the A-body musclecars, they didn't build the famous 383 and 440 Darts that the dealership is associated with, as was often thought. Other than the original prototypes, the big-inch Darts were built by Chrysler or farmed out to Hurst Performance Research. It says something about Grand Spaulding's reputation among the performance faithful that people have naturally assumed Grand Spaulding, rather than the factory, built the 440 Darts.

Conjuring Up Demons

While some Dodge dealers looked upon the success of Plymouth's Duster with envy, the arrival of the sporty fastback at Plymouth dealerships was no big deal to him, according to Kraus. "We always did our own thing. We didn't pay too much attention to what they were doing," he said.

"Don't forget, they gave us the 1970 Challenger," Kraus remembered. "We took the 1970 Challenger, and put different GSS packages on it. They could buy it in a half-dozen different ways—which actually knocked the Duster for a loop. All the guys thought that we got special Challengers. We naturally started advertising these Challengers in all the performance magazines, and *Drag News*, and we were on the air every night from 12 to 5 in the morning on WLS, that went through 50 states. We put on there the GSS Challenger. That's how we handled it."

Not that Kraus looked down his nose at Dodge's version of the Duster when it arrived. On the contrary, when the 1971 Demon was introduced, the Dodge reps brought a devil's costume to Grand Spaulding Dodge, which Kraus wore, even going out into the middle of Grand Avenue to catch people's attention. "They tied it into the devil and Mr. Norm and performance. It was good advertising. We ran with it," Kraus said.

After a one-year diversion to the E-body Challenger, Mr. Norm's special GSS performance car returned to the A-body platform. Rather than working with a big-block V-8, however, this time Grand Spaulding's GSS Demons were powered by the 340-ci version of Chrysler's small-block engine family.

The 340 was already plenty hot as a performance engine, but a new development from the 1970 models seemed a perfect match for the GSS Demon. Dodge, trying to make the most of its involvement with the

The 1972 supercharged Grand Spaulding Demon GSS sold for $3,595, compared to the Demon 340's base price of $2,759.

A Mr. Norm's Demon GSS was not known for its interior appointments—the money went into the drivetrain. Even this car was sold with the base three-speed manual transmission. But by ordering the factory options, a customer could create a comfortable automobile.

Sports Car Club of America's Trans-Am racing series, had released a special T/A Challenger in 1970 (as had Plymouth, with its AAR 'Cuda). The T/A had many unique characteristics to distinguish it from the R/T model, but the most intriguing was its Six-Pack induction system. Employing a trio of Holley two-barrel carburetors had worked wonders for the 440 Magnum V-8 in 1969, nearly elevating it to the level of the famed 426 Hemi. The system worked similar wonders for the 340.

For use on the 340, Dodge had rated the Six-Pack engine at 290 horsepower. As with the standard 340 four-barrel, that was a fancifully low rating.

The 1971 Grand Spaulding Demon GSS 340 Six-Pack was advertised at $3,295. (The 1971 Demon 340's base price was $2,721.) Among its listed features were the three Holley carbs atop the high-rise Edelbrock intake manifold; the special Grand Spaulding traction bar; mechanical, progressive carburetor linkage to replace the vacuum linkage; 3.91:1 rear end gear ratio; aluminum valve spring retainers; a low-profile aluminum Six-Pack air cleaner; a competition fuel pump; and high-volume oil pump.

As part of the Grand Spaulding power tuning, the Six-Pack carbs were rejetted and calibrated and the distributor was recurved. From that starting point, a buyer could keep adding performance gear as long as his wallet held out.

Super Stock & Drag Illustrated magazine got its hands on one of the early Six-Pack GSS Demons and came away impressed. "On the expressways, on-ramps became launching pads as we experienced straight-linkage, six-barrel operation in low gear and second, leaving a pair of black stripes on the concrete for a good many feet before backing out and easing into traffic," they gushed. "The Demon 340, with twin scoops and spoiler not really giving secrets away, is a light car, and it drives lightly, with little effort required to change lanes and turn corners. And then, too, the modifications made by Grand Spaulding add little or no bulk to the car, while they add power."

Fitting the Six-Pack induction to the Demon's 340 V-8 was an obvious step, about as natural as falling out of bed, given that most of the triple-carb pieces were factory-available Mopar parts. But topping that for 1972 required a bit more imagination.

"We said 'What are we going to do for next year?'" Kraus recalled. Inspiration proved pretty close to home—Grand Spaulding Dodge had been sponsoring supercharged race cars with Gary Dyer at the wheel for years. Among race fans and car lovers in general, few pieces of high-performance equipment pack the same thrill as a blower peeking from underhood. If that high-speed image could be transferred to an affordable street car, buyers might potentially line up around the block. "Why not see if we can get a supercharger made for the street?" Kraus concluded at the time.

Of course, for practical purposes, a sky-scraping Roots blower poking out of the hood was not a realistic solution, but the California-based Paxton Superchargers made a centrifugal unit that fit nicely under stock hoods. And there was a precedent for a centrifugal-supercharged musclecar. Carroll Shelby himself had offered the option on his 1966 GT-350 Mustangs.

For once, emission control regulation worked in favor of a performance car. "They were already lowering compression at that time, which just worked out perfect for the supercharger," Kraus said.

Although the Grand Spaulding team had to work through a couple teething problems with the 1972 GSS—such as floats that collapsed inside the carburetor—overall Kraus recalled that creating the supercharged 1972 Demon GSS package together was remarkably straightforward. "Putting a supercharger on is no big deal," he said.

The price of admission was higher for the supercharged cars than for the 1971 Six-Pack Demon. The 1972 GSS started at $3,595. The listed equipment included the Paxton supercharger, a pressure box around the carburetor, "full-capacity hoses" and a special "fresh-air intake and filter unit," a modified fuel pump, a fuel pressure regulator, competition oil pump package, oversize pulleys, aluminum valve spring retainers, and Grand Spaulding's dyno-tuning. About six months after production began, the dealer began fitting the Paxton with a smaller blower pulley, yielding higher boost.

The standard supercharged Demon GSS transmission was the heavy-duty TorqueFlite automatic. Customers could order a supercharged GSS with a four-speed transmission, but Mr. Norm didn't recommend it because the TorqueFlite better matched the supercharged 340's power curve.

Kraus recalled that the supercharged cars sold well, and surprised a few people along the way. "Paxton was calling me," Kraus said. "They said, 'How is it that your Demon is running in the high-12s, low-13s?'

"I said 'I don't know. Maybe it's the Midwest air!'"

Crosswise With the Law

Looking back over his GSS 383 Darts, his GSS 440 Darts, and his Challenger and Demon GSSs, Kraus used different definitions in considering his most successful car.

"Successful means sales," he said. "There are two classes as far as sales are concerned. Capturing the imagination of the public was the 1972 supercharger. As far as selling numbers, it was the Dart 440."

Performance-minded dealerships like Grand Spaulding Dodge flourished for a time, but were almost uniformly out of the performance business by the early 1970s. When the musclecar end of the business slowed down, thanks largely to rising insurance costs to buyers and a shift by the Big Three away from high performance, these dealers drifted off to focus on luxury conversions, or custom vans, or wherever the next hot segment led. An Environmental Protection Agency crackdown of dealers who discarded the factory emission controls killed the remaining hot rod-minded dealers.

"We didn't do anything in 1973 because the word got out with the emission control," Kraus said. "We had calls coming into the place and you knew it was the government. Just the way they acted. We said to ourselves we've got to get into a different area. And Chrysler already told us they were not going to come out with any more performance cars." After 1972 Grand Spaulding made its money through fleet sales and trick vans.

Still, those memories linger. Mopar fans have lovingly restored the A-body GSS cars during the last couple musclecar nostalgia waves, and a documented Mr. Norm's model is worth a substantial premium over the same musclecar in basic, factory trim. A revived Mr. Norm's Sport Club was even established in 1999, with a new club magazine, The Broadcast Sheet, to spread the word.

"We were lucky to be in the right place at the right time," Kraus said. "And don't think we didn't take advantage of everything."

PLYMOUTH
OHIO
DCH-067
AUGLAIZE
DAYTONA
RADIAL

five

Dart Sport 360 & Duster 360

1974–1976

At a time when American musclecars were dropping away like so many buffalo on a nineteenth-century American prairie, the Duster 360 and Dart Sport 360 kept the high-performance spirit alive. The two cars offered more bang for the buck than any other American performance car. The Pontiac Trans Am might have been the undisputed top American performance car in the mid-1970s, but for the extra money, it should have been. A Duster 360 could be purchased for roughly $800 less, with insurance costs to match, and a quarter-mile ET only a tick behind and sometimes ahead, depending on equipment.

Prior to 1974, the 360 had primarily been a mildly tuned family car and truck engine. Introduced in 1971, the 360 was not crammed full of the type of high-performance hardware in the 340. Fitted with a two-barrel carburetor, this Polara-mover was rated at 255 horsepower. When the redesigned 1972 Dodge pickups were released, the 360 proved to be a perfect midsized engine option for the truck.

The 340, meanwhile, had been designed almost exclusively for performance, and was therefore more expensive to produce and harder to get to pass ever-tightening emissions standards. With its forged steel crank, Hemi timing chain, and big-valve heads, the 340 was definitely a product of the tire-smokin' 1960s. However, even the 340 had been slowly emasculated in the early 1970s. The engine received the pedestrian 360 two-barrel heads in 1972, and lost its forged steel crankshaft in 1973, replaced by a cast crank.

As if emissions regulations and high insurance rates hadn't done enough to kill the American

The 1974 Duster was not substantially changed from the 1973 models, although the familiar 340 V-8 was replaced by a 360-ci small-block.

musclecar, by 1974 the country had suffered its first Arab oil embargo. High-performance engines like the 426 Hemi and 440 Six-Pack had been flung from the lifeboat in 1971 and 1972 respectively, and keeping both a watered-down 340 V-8 and the go-to-work 360 in the line-up made little sense. In the 1970s market, the 360 was the more appropriate engine, being suitable for both regular passenger car and pickup use, as well as performance.

The 360, then, starting in 1974, became both a performance engine for the Duster and Dart Sport, and the two-barrel everyman engine it had always been. The Duster 340 model was simply changed to Duster 360, likewise with the Dart Sport.

With a four-barrel carburetor, the 1974 360-ci V-8 was rated at 245 horsepower, 5 more than the late 340 had produced in its final year. With the extra displacement, torque was up as well, from 290 ft-lb to 320. The 360 V-8 proved more than sufficient to make the lightweight Duster and Dart Sport among the top performers of the mid-1970s. The Duster 360 was still one of the best performance bargains of its time. Its factory A.D.P. (manufacturer's suggested advertised delivered price) listed at $3,288. The Dart Sport 360 started slightly higher, at $3,320. At $3,059, the similarly sized Chevrolet Nova SS was the duo's most obvious competition. Its 350-ci V-8 put it close to the Duster/Dart in performance, although not quite ahead.

The 340-ci small-block was stroked out to 360 cubic inches for 1974. While other automakers' engines were shrinking and growing weaker in the emission-controlled 1970s, the 360 still packed a punch. The Duster's 360 V-8 was rated at 245 net horsepower in 1974.

Cars magazine conducted a comparison test of a Dart Sport 360 and a Chevy Nova SS 350. "While the Nova SS delivered what could be called brisk acceleration, the 360 Dart was loose enough to actually snap you back in your seat," writer Joe Oldham noted.

With automatic transmission and 3.55 rear axle, the Cars testers managed a 14.68 quarter-mile at 92 miles per hour. The Nova, with its automatic and 3.42 axle ratio, turned a 15.14 at 88 miles per hour.

A sub-15-second quarter-mile was a serious achievement from an American car in 1974, but reality intruded even here. "One of the problems in

"Tuff" steering wheel and traditional Mopar console kept interior appointments for the 1974 Duster 360 familiar to Mopar enthusiasts. The console was a $54 option that required ordering the $127 bucket seats as well.

driving the car at its limit was the engine," Oldham wrote. "At low revs, the smog controls effectively cut off quick response. Then as we pushed down more on the throttle, the power would come on with a rush and we'd be fighting to control the power oversteer. No fault of Chrysler. It's the fault of the feds and their smog stuff."

If Mopar management was paring down the engine choices, it was also trying to eliminate other duplication in the ranks. In 1974 it shifted the regular four-door Valiant to the Scamp's 111-inch wheelbase, simplifying matters. Since 1971 the Valiant line had included a four-door Valiant, a Duster, and the Dart-based Scamp, with the Duster and Valiant four-door built on a 108-inch wheelbase. The Valiant had always had a shorter wheelbase than the Dart, but the Scamp hardtop, for reasons of expediency, had been built on the Dart chassis.

Of course, for every simple step forward for Chrysler in the 1970s, there was seemingly one step back, or at least sideways. The newest Valiant for 1974 was the Brougham, which meant that Plymouth was suddenly making a move to a market no one knew existed. The Dart had always been marketed as a "big compact," and now Plymouth was in the business of building posh economy cars.

The Brougham model came standard with the larger 225 Slant Six, with V-8 optional. But powertrains didn't matter—the Brougham's purpose was luxury, and to that end it came with every pampering feature Plymouth could bolt on. The Brougham was by far the most expensive Valiant in 1974, with prices starting at $3,794 for the two-door hardtop, and $3,819 for the four-door sedan. By comparison, the Duster 360's base price was $3,288, and the next most expensive model, the Scamp, started at $3,017.

General Motors in general, and Pontiac in particular, exhibited a strong influence on the struggling Chrysler Corporation. This Duster proposal from Plymouth stylist Fred Schimmel shows a clear Pontiac influence. "They were very strong on a personal luxury car," Schimmel said. "For several years there, that was the thing—'what can we do to have a Grand Prix or Monte Carlo out of every car line?' Besides the muscle version, they wanted the personal luxury version."

The Dodge answer to the Valiant Brougham was the Dart Special Edition. The Special Edition came with a padded vinyl top, color-keyed hubcaps, special cloth seating, plus what luxury features existed on the lower-line Darts. Special Edition prices for V-8 models started at $3,945, about as much as a well-equipped Charger SE model.

The Dodge boys were tacking in several different directions at once, though. Their new attention grabber for 1974 was the Hang Ten edition. Although the surfing craze had peaked in the 1960s, the surfing image was still very much a marketable phenomenon. The Hang Ten footprints logo could be found on everything from shirts to accessories.

On the Dart, the Hang Ten package featured "wave crest" tape stripes, orange shag carpeting, including orange carpet on the rear of the fold-down seat, white vinyl bucket seats with special woven vinyl inserts, and orange-and-white touches on the dash and console.

Richard D. McLaughlin, Dodge general sales manager, explained the Hang Ten's appeal. "The Hang Ten is the result of a more practical approach to show car design, which can be applied to current cars that people actually own, instead of futuristic 'dream' models which are pretty to look at, but which rarely become reality," he said. "Our design department is responding with cars and trucks color-keyed and oriented to personal interests, hobbies, or topical themes."

The importance of the Dart, Valiant, and Duster to Chrysler Corporation became even more pronounced in 1975. In March 1974, the E-body Dodge Challenger and Plymouth Barracuda were canceled, victims of low sales. Every bad thing that could possibly happen to the high-performance

This Plymouth A-body proposal from the 1970s has a resemblance to American Motors' Hornet. "This is a full-size clay model, where they built a buck that you put all the clay on," recalled Plymouth stylist Fred Schimmel. "It isn't solid clay, they build up a box inside of there—and after they refine the surfaces, the model is covered with Dynoc." Dynoc is an adhesive covering not unlike shelf paper, useful because it stretches well. "They could paint it and you could stretch it over all these contours," Schimmel said.

ponycars fell from the sky—a late start, prohibitive insurance rates for customers, federal regulations that hurt performance cars the most, and an oil embargo that turned musclecars into "gas guzzlers" almost overnight.

With the E-bodies put to sleep, the Duster and Dart Sport were for all intents and purposes the sole youth-oriented sporty cars available from Chrysler Corporation, unless one wants to count the Mitsubishi-built Dodge Colt. The Charger, once the flagship of the Dodge Scat Pack, had devolved into a squishy, overweight Cordoba clone. Road Runner graphics were plastered onto the similarly bulky Fury, but with the standard engine reduced to the 318-ci V-8, few people cared.

Not surprisingly then, the 1975 Dart line-up consisted of a wide range of models. Buyers had to choose between the Dart, Dart Sport, Dart Swinger, Dart Swinger Special, Dart Custom, Dart Sport 360, and Dart Special Edition. At Plymouth the Valiant line was divided into Valiant, Valiant Custom, Valiant Brougham, Scamp, Duster, Duster Custom, and Duster 360 models.

The 1975 models were given new grille treatments to keep the line fresh, but also a few functional improvements, such as radial tires and a new

"Overdrive-4" manual transmission. Additional sound deadener was inserted into the cars, and an exhaust resonator helped lower the decibel level. The Dart Sport 360 and Duster continued, with the engine downrated to 230 horsepower for the new year.

A Premature End

The Dart and Valiant had been two of the few bright spots for Chrysler Corporation in the 1970s. Between the two they routinely captured a third of the compact market in the United States. They represented nearly half the corporation's domestic car sales, each capturing approximately a quarter-million buyers a year. With that sustained popularity there was no reason to restyle or re-engineer the cars with any frequency, keeping the corporation's investment low and profits steady.

Meanwhile, other segments of the Chrysler product line were proving to be a drag on the bottom line. The Dodge Coronet and Monaco, and Plymouth Fury and Grand Fury, were large, invisibly styled, heavy, slow, gas hogs at a time when the public was searching for values at the other end of the market. The A-body cars, along with pickup trucks and vans, were doing most of the financial heavy lifting at Chrysler Corporation.

But no car can hang around the market forever. Chrysler had been working on the replacements for the two throughout the mid-1970s. Although similar in size and shape, the replacements had altogether different personalities. The Dodge version was named Aspen, the Plymouth model the Volare. They were designated the F-body cars. Their strengths included a revised suspension that offered a superior ride compared to the older A-bodies, but the car came with a trunkload of weaknesses.

Although not unattractive cars, the Aspen and Volare exhibited the sort of split personality common in so many American cars in the 1970s. If performance and style had been the hallmarks of 1960s automotive design, the two prevailing philosophies in the 1970s were economy and luxury. If the two sound at odds with each other, it's because they were.

The Arab oil embargo of 1973 and horrendous inflation throughout the decade forced the issue of fuel economy on most Americans, but sales of "personal luxury" cars were also high as the baby boom generation took the next step up life's ladder. Sales of Hondas, Toyotas, and Datsuns exploded, and so did sales of the Chevrolet Monte Carlo, Pontiac Grand Prix, Oldsmobile's Cutlass, and even Chrysler's own Cordoba.

Every automaker tried to tack luxury styling cues onto its various models. Once stoic designs sprouted stand-up hood ornaments and names like "Brougham." And so it was that the Aspen and Volare, successors to the sensible Dart and Valiant,

This handsome Duster proposal from Plymouth stylist Fred Schimmel was dated 1970. "This was the next body of A-body," he recalled. "This is just a rear end proposal that was trying to keep a very flowing, aerodynamic shape to the back end." Schimmel said. "This was developed as a two-door, and maybe that's where we were starting—with a two-door Duster, and then developing a four-door Valiant out of that." Schimmel worked at the Plymouth styling studios from 1955 to 1975, most of the time working on Valiant and Barracuda designs under studio chiefs Dick McAdam and Jerald Thorley. In 1971 he moved to the advanced design studio.

came to market with fussy, overwrought, upright grilles, Landau vinyl tops, and opera windows. Even the names were pretentious, suggesting mountain vacation homes and international jet-setting. In keeping with their new position, the new cars were more expensive.

But that was the least of the cars' problems. Lee Iaccoca took over as Chrysler CEO in 1978 and had the unenviable task of saving a failing automaker. The Aspen and Volare had been a big source of the company's problems. As he recalled in his 1984 autobiography, *Iacocca, an Autobiography*, "The company had also run into big problems with quality. Among the worst examples were the Aspen and Volare, the successors to the highly acclaimed Dart and Valiant. The Dart and Valiant ran forever, and they never should have been dropped. Instead they had been replaced by cars that often started to come apart after only a year or two," Iacocca wrote.

"Aspen and Volare were introduced in 1975, but they should have been delayed a full six months. The company was hungry for cash, and this time Chrysler didn't honor the normal cycle of designing, testing, and building an automobile. The customers who bought Aspens and Volares in 1975 were actually acting as Chrysler's development engineers."

Granted, the Dart, and especially the Valiant, had a reputation for being grandma's preferred mode of transport, several years worth of GT and GTS models notwithstanding. The A-body line's image could hardly be described as glamorous. But the cars also had a reputation for being bulletproof, and that kind of reputation takes years to build and hardly any time to destroy. The Aspen and Volare took that hard-earned reputation, wadded it up, and blew it out a poorly fitting tail pipe.

"The Aspen and Volare simply weren't well made," Iacocca related. "The engines would stall when you stepped on the gas. The brakes would fail. The hoods would fly open. Customers complained, and more than three and a half million cars were brought back to the dealer for free repairs—free to the customer, that is. Chrysler had to foot the bill."

The 1975 Dart Sport could be ordered with the full line of engines, from Slant Six, to 318 V-8, to Dart Sport 360. One popular option was the manually operated sunroof, which, when teamed with the fold-down rear seat feature, was known as the "Convertriple" package. *Reprinted with the permission of the DaimlerChrysler Corporate Historical Collection*

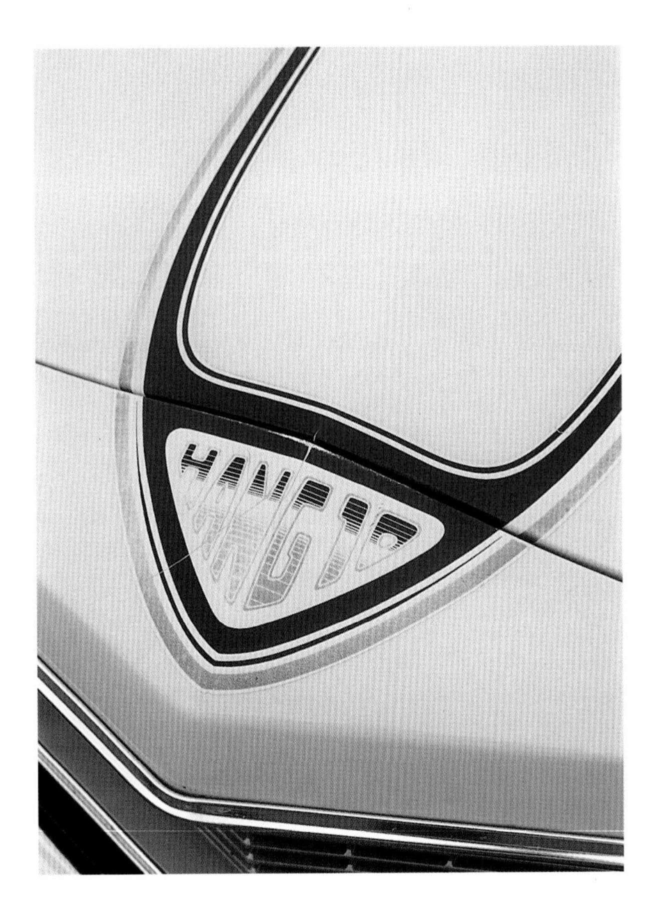

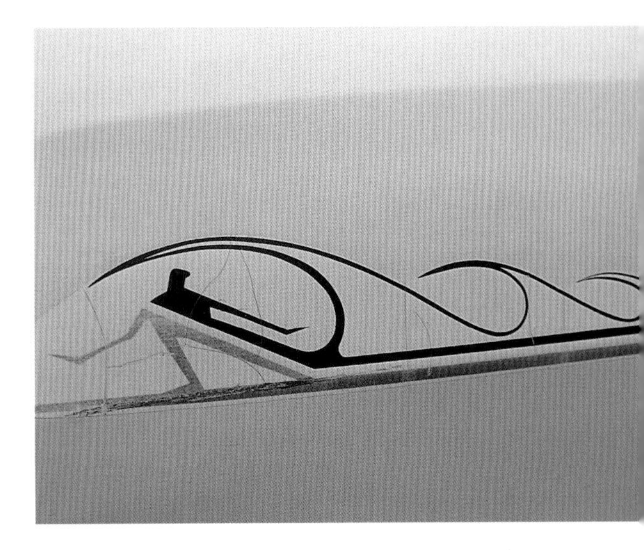

In theory, surfer dudes and surfer wannabes would be drawn to the specially tailored Dart Sport with Hang Ten package. In actuality, the Hang Ten Dart Sport wouldn't be a bad choice for wave enthusiasts, since the fold-down rear seat opened up 6 1/2 feet of cargo space, enough for a smaller surfboard. Getting sand out of the Hang Ten's orange shag carpet could be a pain, though.

Fortunately, car buyers in 1976 still had a choice of economical Dodges and Plymouths. The new compacts were sold alongside the existing cars in 1976. And in their last year in the U.S. market, the Dart, Valiant, and Duster were pushed more than ever toward the economy car market.

The 1976 Dart models consisted of the Dart, Dart Sport, Dart Swinger, and Dart Swinger Special. A Dart Lite package for the Dart Sport model was intriguing. On the Lite, the now venerable 225-ci Slant Six used an aluminum intake manifold, and the engine block was lightened to save weight. The Dart Lite also used an aluminum hood, trunk lid inner panels, and bumper reinforcements. If the Overdrive-4 transmission was ordered with the Lite, the casing was made of aluminum. The Dart Lite weighed 150 pounds less than the standard Dart.

Economy-think was everywhere in 1976. While the Dart Sport with 360 four-barrel was still available, it was fitted with a highway-loafing 2.94:1 rear axle ratio, with only a 3.21:1 optional. The 49-state 318 V-8 came with an incredibly high 2.45:1 axle ratio.

R. D. McLaughlin, vice president of Chrysler's Automotive Sales Division, acknowledged the Dart's importance in a 1976 press release. "In a price-conscious market, Dart will have an advantage," he said. "The car has more than a decade of building a strong reputation for extra value as a new car purchase and good value at resale. These are reasons why we will continue to market the Dart while introducing the new compact Aspen."

At Plymouth, the 1976 Valiant line was cut back to make way for Volare. The remaining models included the Valiant, Scamp, Scamp Special, and Duster. The Custom, Brougham, and Silver edition were the major option packages. The Silver Duster was basically a trim and appearance package, with a silver exterior and red interior color scheme supplemented by unique body-side striping. It took the place of the Gold Duster, which made its final appearance in 1975. The Silver trim package was a $178 option. The aluminum-intensive, lightweight twin to the Dart Lite was called the Feather Duster, showing that someone at Chrysler still had a sense of humor.

The Duster 360 model disappeared, although the engine was still available for $392. The 360 V-8 was offered with the TorqueFlite automatic transmission only. For emissions reasons, the 360 was not available in California. The 318 V-8 continued to be the bread-and-butter engine for the A-body Plymouths.

Although both the Dart Sport and Dusters were available with 360-ci V-8s in 1976, most of the promotional push went behind the new Aspen and Volare "performance" models, the Aspen R/T and Volare Road Runner. The two were bestowed with great performance names from the past, but as performance

Worldwide Fans

Perhaps more than many American cars, the Dart and Valiant are popular throughout the world. The same virtues that made the A-bodies popular in the United States—low cost, economical operation, sturdy construction, functional styling—were even greater selling points in other countries. The Slant Six engine was a natural for international markets, offering good power and economy and, thanks to its longevity, good parts interchangeability through the years.

Naturally, as closely tied as the two countries are, Darts and Valiants were a common sight in Canada. Initially, only the Valiant name was affixed to the Canadian-built cars. The early Canadian Valiants were odd combinations of Dart and Valiant sheet metal. Later in the 1960s Canadians could choose between cars with Dart or Valiant sheet metal, although all were labeled as Valiants. Starting in 1967, Darts and Valiants were marketed separately in Canada, and were more closely aligned with their American-market cousins.

It's doubtful any countries embraced the Chrysler A-bodies as strongly as did Australia and New Zealand. Before Chrysler pulled out of Australia at the end of the 1970s, the Valiant was a huge seller in that market. As with its first year in the United States, the Valiant was marketed as a separate brand in Australia. The Valiant platform later expanded to include a two-door, fastback Charger model, a Drifter model, and the El Camino-like Utility truck.

Valiants were assembled in Australia starting in 1962. Besides serving the Australian market, the cars were exported to New Zealand, South Africa, and other United Kingdom outposts. In 1963 production was expanded to include the AP5 model. This Aussie-market car looked like the American-market Valiant, but had its own unique trim and roofline.

Just as the American musclecar mania of the 1960s trickled down to the compact economy cars, the Australian Valiant spawned its own performance versions. One of the better known of these was the Pacer, introduced in 1969. The Pacer came with a high-performance version of the Slant Six, a four-speed transmission, and heavy-duty suspension.

The Charger, a short-wheelbase, two-door sport model, was introduced in 1971 on the A-body platform. Some things translated perfectly to overseas markets, and the R/T name as a designation for performance was one of them. The Australian musclecars adopted all of the Chrysler performance jargon, with the top Charger R/T having a "Hemi Six-Pack" underhood.

The Australian Hemi did indeed have hemispherical combustion chambers, but only six of them. The 265 Hemi was an inline six created for the Australian market. The "Six-Pack" referred to the trio of two-barrel Weber carburetors that fed the Hemi. The 1972 Charger finally received a four-speed transmission, and with the 265 Hemi's power up to 302 horses, the 1972 Charger Hemi Six-Packs are considered among the quickest cars, and perhaps the quickest, ever built in Australia. These cars found great racing success in New Zealand's Production and Endurance racing classes. The 340 V-8 also found its way under Charger hoods in 1972.

The intriguing Australian Hemi Six-Pack might have made a great engine for the U.S. market in the 1970s, packaged correctly. But it's understandable why it was not brought in. There was a precedent for how well high-performance sixes sold in the United States. Pontiac had introduced the wonderful Sprint Six in the Tempest in 1966, and in the Firebird in 1967. With its overhead cam and four-barrel carb, the Sprint Six produced 207 horsepower in 1966 and 215 in 1967. But Americans, raised on neck-snapping V-8s, never warmed up to the engine. The Sprint Six never sold in the kind of numbers GM had hoped. It's likely the Hemi Six-Pack would have had similar difficulties, although the post-oil-embargo market might have been friendlier. But then again, trying to get U.S. emissions certification for a trio of Weber carburetors would have been a neat trick.

South Americans knew the A-body through the Dodge Dart name, rather than Valiant. The Dart was assembled in Colombia, Venezuela, Peru, and Brazil, and was found throughout Latin America. Besides the Dart, the South American spin-offs from the A-platform included the Polara, Polara R/T, Coronado, and GTX. Variations of the Dart were sold in South America through 1982. No doubt some in North America wished the same was true.

The Duster 360 was given a new tape stripe treatment for 1975, along with minor grille and taillamp alterations. Power brakes were standard on the Duster 360. A sunroof remained an option. *Reprinted with the permission of the DaimlerChrysler Corporate Historical Collection*

cars, they left something to be desired. The days of sub-15-second quarter-miles were long gone. Hell, even sub-16-second quarter-mile ETs were only dreams from another time.

Engine choices for the 1976 Road Runner and Aspen R/T included either the 150-horsepower 318 or the 170-horsepower 360 V-8. The Road Runner was easily identified by its body-length, multicolored stripe and cartoon character logo. The Aspen's R/T identification at least bore a close relationship to the Mopar musclecars of the past, keeping the same familiar logo used in the 1960s. A wild deviation was the Aspen Super Coupe, introduced in 1978. The Super Coupe, with its large, tacked-on fender flares, fat quarter-window louvers, and multicolored striping, was definitely a creature of the 1970s.

The Aspen and Volare remained in production through the 1980 model year. They were the direct replacements for the Dart and Valiant, but the true spiritual heirs to the two may have been the K-cars. The Dodge Aries and Plymouth Reliant, front-wheel drive compacts introduced for the 1981 model year, were closer in concept and feel to the original A-bodies from the 1960s. These blocky economy cars had the simple styling, low-price, and fuel-sipping abilities of the earlier cars. Yet thanks to front-wheel drive and overhead cam four-cylinder engines, they were also completely contemporary in their time, not just the same old car with new sheet metal. The compact, unibody 1960 Valiant, with its brand-new Slant Six engine, was also a positive break from the flathead, body-on-frame past.

Just as the Dart and Valiant gave birth to the Barracuda, Duster, and Demon, the K-car chassis spawned all manner of economical off-shoots. The Aries and Reliant gave way to the Dodge 400, Lancer and Shadow, the Chrysler Le Baron, and a variety of Plymouth economy cars.

The legacy of the muscular Darts and Dusters continues to occasionally exert itself, although sometimes just to mine the brand equity tied up in the nameplates. The Duster name was revived briefly in 1993 on the Plymouth Sundance, one of the last of the K-car spin-offs. It's doubtful they'll rate their own book someday. Now, a 340-powered Sundance, with bumble bee stripes and a big Hurst shifter, that would be one for the history books.

six
Competition Darts

It looks odd staring back at you from the pages of the Sports Car Club of America's (SCCA) record book, but there it is—the Dodge Dart was the most successful Mopar in SCCA Trans-Am racing during the height of the musclecar era. The Dart scored more wins than the Barracuda (one), and a lot more than the famed T/A Challenger (zero). Of course, the Dart's victory total was meager, with only three wins, but still, it's an oddity. The pedestrian Dart was actually a more successful road racer than its famous ponycar brethren.

The architects of that success were Bob Tullius and Tony Adamowicz. The two started in Trans-Am at the ground floor, which helped the Dart establish itself before the onslaught of factory-backed teams made the series a Mustang/Camaro showcase. They finished second overall, first in the over-2-liter class, at the inaugural Trans-Am race at Sebring, then went on to win later in the year at Marlboro. Tullius finished fourth in points for the season, with Adamowicz finishing fifth. In 1966 Tullius won the season opener at Daytona, the final time a Dart would visit the Trans-Am victory lane. The Mustang and Camaro waves were approaching.

Although road-racing Darts hardly fit the image of Chrysler's economical compacts, in truth the cars were raced from the beginning. Prior to the Trans-Am series, the Valiant made a splash with its 148-horsepower, 170-ci Hyper-Pak six. In NASCAR's Compact Class road race at Daytona in 1960, likely the Valiant's first competitive event, the cars finished first through seventh.

Although best known for offering economical transportation, the Dart, Valiant, and Duster always made good raw material for racing. They were

Perhaps the most feared and capable of the Super Stock factory race cars built by Dodge was the Hemi Dart. Built in 1968 only, the Hemi Darts were put together at Hurst Performance Research. The 426 Hemi V-8 graced the engine compartment of thousands of Dodges, but was most effective in the compact Dart. Fiberglass fenders and hood helped reduce weight to the bare minimum. *Photo courtesy Rob Reaser*

The 1968 Super Stock Darts were fitted with the 12.5:1 compression race Hemi. This version of the engine featured dual four-barrels on an aluminum cross-ram intake. *Photo courtesy Rob Reaser*

cheap, lightweight, and offered a variety of V-8 engines. Like the majority of musclecars, however, the Dart and Duster won most of their trophies at the drag strip.

Drag racers began modifying the Slant Six for quarter-mile competition from the start. The Slant Six was a good fit in the early Modified Production classes, and was at home in both E/MP and F/MP. One advantage was that few automakers bothered to develop high-performance parts for their six-cylinder engines, as Chrysler had done with the Hyper-Pak modifications. Slant Six racers had a huge advantage thanks to the factory-engineered pieces.

The Hyper-Pak for the Slant Six incorporated traditional hot rodding techniques that worked well with the 170-ci engine. The stock one-barrel carburetor and intake manifold were ditched in favor of a four-barrel carb and long-runner intake. The exhaust manifold was a header-type design. The compression ratio was raised to 10.5:1, and a hotter 276-degree cam and stiffer valve springs helped bump output to 148 horsepower. The package also included a heavy-duty clutch to help keep the drivetrain intact.

Modified Production was specifically designed for customized cars that were still driven on the street, and was ideal for low-budget racing efforts. At the other end of the money scale was the Factory Experimental division, but even here the National Hot Rod Association (NHRA) made room for six-cylinder cars, classifying them as C/FX.

One of the earliest A-body believers was Ron Root. He raced a 1963 C/FX Dart to great success, and won the 1964 NHRA Winternationals Street Eliminator championship. Pete McNicholl was another Dart achiever, scoring a number of victories in his high-13-second F/MP Dart.

The Dart's and Valiant's appeal to racers increased in 1964 with the addition of the 273-ci V-8 to the option list. When the high-performance 235-horsepower 273 V-8 was released the following year, a racing Dart was no longer an oddity—it was a junior musclecar. The V-8 A-body cars usually fell into the C/MP and D/MP classes.

In 1966 the factory helped Dart racers along even further with the introduction of the V-8 "D/Dart" hardtop, so named because the car was tweaked to meet D/Stock specifications of the NHRA, AHRA, and NASCAR. The D/Darts were regular production line vehicles, although like their race Hemi and Max Wedge counterparts, they came with the warning that, "Due to the expected use of these vehicles, no warranty coverage applies."

Although technically still street cars, the D/Darts were highly tuned vehicles, and definitely not the type of street car the average cop enjoyed seeing tooling around town. The power-to-weight ratio was strong, with 2,946 pounds being dragged around by a 275-horsepower 273 V-8.

The extra 40 horses came by way of a modified Holley 4160 four-barrel carburetor with low-restriction air cleaner, a distributor altered for quick advance, a more radical Camcraft camshaft with .495-inch lift on the intake side, .505-inch lift on the exhaust, and 284 degrees duration, Racer Brown valve springs, a modified intake manifold, and Doug's headers feeding a large single exhaust system.

The Super Stock Hemi Darts were stripped of many interior features to save weight. The window regulators were removed and replaced with pull-down straps, and the low-back seats, with drilled mounting brackets, were designed for the A-100 Van. Windows were Plexiglas. *Photo courtesy Rob Reaser*

The rest of the car was race-ready as well, with a Sure-Grip 8 3/4-inch rear end and 4.86:1 gears, a four-speed transmission with a Weber heavy-duty clutch, a heavy-duty suspension, and 6.95x14 tires. Chrysler also offered the parts separately for racers wishing to build a D/Dart at their own pace.

Car Craft magazine tested a D/Dart in a 1966 issue and recorded a 14.33-second quarter-mile at 94.21 miles per hour—and that was in as-shipped-from-the-factory condition, with only Cragar S/S wheels and Eliminator Prowler street slicks for prep.

Writer Dick Scritchfield noted, "The engine still had less than 150 miles on it, so the potential was definitely there. With a super drag strip tune and a few alterations in the chassis department, the D/Dart should make the record lists."

The 273 was an improvement over what had come before, but the 340-ci small-block introduced in 1968 was an even bigger improvement—on the street, at least. The 340 V-8 was another in a long line of musclecar engines with a sandbagged power rating. The 340 was rated by Chrysler Corporation at 275 horsepower, but the paper number didn't fool the NHRA—the 340 was factored to 310 horsepower for class competition. Although seemingly ideally suited for the drag strip, the 340 was often handicapped by the unfavorable factoring.

Big Ideas

Although the Dart and its A-body cousin, the Barracuda, were fine racing vehicles if confined to their own niche, the cars did suffer a few handicaps. One A-body drawback was the unusual 4-inch diameter bolt circle Chrysler used on the hubs. While fine for economy car use, the unconventional pattern was an annoyance to racers, who found their wheel choices limited. These drivers had a hard time fitting their cars with the larger 15-inch wheel and slick combinations preferred for racing. (Starting in 1973 the disc-brake-equipped Darts and Dusters were

Lightweight and potent small-block engines have made the Dart a natural for grassroots drag racing competition.

switched to a bolt circle diameter of 4 1/2 inches.) Also, the less-than-generous size of the wheelwells on most A-bodies made fitting large, grippy tires an exercise in wishful thinking.

Dart and Valiant enthusiasts were able to race in the less glamorous economy car classes, but with only six-cylinder and small V-8 engine choices, the A-bodies would never race in the top classes, never be the main event. Until, of course, racers found ways to stuff Chrysler's monster engines into the corporation's smallest cars.

That came to pass in 1967.

With the redesign of the 1967 Dart, Valiant, and Barracuda, came additional room in the engine compartment. The 1967 Darts rode on a 2-inch wider front track, and were 1 inch wider overall. They were, as Dodge liked to remind potential customers in advertisements, the largest compact on the market.

Even with the extra size, though, Chrysler didn't rush to give its compacts larger engines. Only the Barracuda started the new model year with the 383 V-8 on its option list. The Dart had to wait a while, and it took some prodding from an outside source for Dodge to follow through.

Dodge dealer Norm Kraus had built his Chicago-area Grand Spaulding Dodge around an image of fast cars and racing success. With racing director and driver Gary Dyer working with him, Kraus' dealership specialized in dyno-tuned performance cars for the street and track. Grand Spaulding Dodge used that notoriety to sell high-performance musclecars by the truckload. The supercharged Funny Cars piloted by Dyer were emblazoned with "Mr. Norm, the high-performance king," lettered along their flanks. The Grand Spaulding match-racers were instantly recognizable at the strip, and attracted young car lovers to Mr. Norm's dealership like a magnet.

For Mr. Norm, a 383 Dart was an obvious step (as detailed in chapter 2). Kraus and crew coaxed the big-block into the engine bay, worked around the steering gear and other impediments, and convinced Dodge that the concept had sales potential. For Grand Spaulding Dodge, the Dart 383 was seen as an image-builder and a way to boost sales among the street performance crowd.

While the big-block Darts were a good starting point from which to build a race car, they needed some help. The first 383-powered Darts and Barracudas were down 45 horsepower compared to the 383 Chargers and Satellites thanks largely to the restrictive exhaust manifolds necessary to make the engine fit. "You can literally pop 30 horses with a set of headers," said Keith Rohm, noted Mopar restorer and car show judge.

Unfortunately for modern restorers, the most common headers available in the 1960s for 383 Darts and Barracudas were fender-well types that required cutting the inner fenders. Many an early Dart was butchered in this manner, although that would hardly be considered a handicap for those wishing to race an older Dart in today's competition.

After the successful creation and acceptance of the 383 Dart GTS, the next step was obvious. If a 383 big-block V-8 would fit, why wouldn't the new 440? True, the 440 came from the Raised Block (RB) side of the big-block engine family, but once the blueprint for a big-block A-body was established, the increased deck height of the 440 wouldn't be much of an issue.

According to Norm Kraus, in late 1967 his Grand Spaulding Dodge dropped a 440 in a Dart and dutifully reported their findings to the factory, although this time the results were hardly a surprise. Their prototype once again was used as a template, this time for a limited run of 1968 and 1969 440 Darts.

As with the 383-ci Dart, to create the 440 cars the driver's side motor mount had to be fabricated, the K-member modified, and the 383 Dart's special driver's side exhaust manifold employed. The 440 cars were built with TorqueFlite automatic transmissions only. Since the 440 crowded out the power brake booster, disc brakes were not an option, but considering the 440-ci Dart's primary use would be driving a quarter-mile at a time in a straight line, it was assumed people could do without.

Dodge contracted with Hurst Performance Research in Michigan to assemble the Grand Spaulding-engineered 440-ci Darts. Hurst was shipped Darts equipped with the 383 engine, TorqueFlite transmissions, and the Sure-Grip rear end for the conversions. Considering the firepower underhood, the 440 Darts were generally very low-key in appearance. Only "440 Four-barrel" plates in the GTS hood bulges gave away the secret.

Mr. Norm's Grand Spaulding operation so successfully marketed the big-block Darts that many people thought they were built by the Chicago dealership. The 440 Darts sold through Mr. Norm were given the special "GSS" treatment, giving them a unique identity. Although the exact number of 440 Darts built is unknown, it is believed several hundred were created. "They sold very well," said Kraus.

Grand Spaulding's work on the big-inch Darts led to a major fix in the traction department. "We had one occasion where we were on the track running one of our Super Stockers one night, and we flipped a driveshaft out of the car. We looked at it and saw what happened. There was a traction bar that was missing off the car. Actually, all it was was a Hemi plate, that came on the Hemi, that sat on top of the differential. We took that, we turned it around, we put a foot-long chrome molly bar on it,

The earliest A-body performance engine was the 170-ci Hyper-Pak Slant Six of 1960 and 1961. Like the 413-ci V-8s used after 1958 in the Chrysler 300 letter series, the Hyper-Pak relied on extraordinarily long intake runners for a "ram tuning" effect. The Hyper-Pak also used a four-barrel carburetor, high-lift cam, header-type exhaust manifolds, and a high-compression ratio. The engine pegged the dynamometer at nearly 150 horsepower. Applying the Hyper-Pak parts to the 225-ci Six could yield nearly 200 horsepower. *Reprinted with the permission of the DaimlerChrysler Corporate Historical Collection*

drilled a hole, and put a snubber plate on the end. Then took about a half-inch off of it. That gave more traction to those Darts than every Mopar product. That's all we did to the Super Stockers."

Car Craft magazine tested a 1969 Dart GSS in a 1969 issue. "With the hood open, all you can see is cubic engine, as this 'baby' occupies all the space from firewall to radiator, and shock tower to shock tower. The fit really isn't that bad, though, as the stock engine compartment sheet metal retains its standard form," the magazine noted. The testers also discovered how difficult it was to translate the 440's massive power output into quick elapsed times. "The stock E-70x14 tires are worthless on this car, as there's no way they can handle the torque this engine puts out," they wrote. "Even so, the initial times through the quarter-mile were quite impressive. How does a 13.71 with a top speed of 105 miles per hour-plus sound?" After installation of Hooker fenderwell headers, 4.30 rear end gears, installation of a pinion snubber, Goodyear slicks, and a supertune, they whittled the ET down to 13.14 at 108.86 miles per hour.

The Dart as Blunt Instrument

With every other engine in the Chrysler fleet finding its way under Dart and Barracuda hoods, it

Dick Landy's Super Stock Dodges were some of the highest-profile Mopars at the drag strip in the 1960s. With his trademark cigar clenched between his teeth, he cut a pretty high profile in person, too. Although he is better known for his Chargers and Coronet R/T race cars, he also piloted Darts, like this 1969 model shown. *Photo courtesy Drag Racing Memories*

was only a matter of time before the inevitable happened—a Hemi-powered A-body. The search for greater performance has always eventually led to fitting the most powerful available engine in the smallest possible car.

And Chrysler had a history of producing factory-built race cars for select drivers. One of the earliest even wore Dart nameplates. The 1962 "Max Wedge" 413-ci Dodge Darts and Polaras, and Plymouth Savoys and Belvederes, were typical of the approach. Take a suitable two-door sedan from the line-up, strip as much weight from it as possible, install the corporation's largest engine tuned for racing, and sell the car to preferred drivers.

In 1962 the Dart was still an entry level version of the standard-size Dodge—although the entire Dodge line-up had been downsized for 1962—and was thus the cheapest, lightest Dodge available that would house a big-block engine. (The smaller Lancer was designed from stem to stern for six-cylinder use.) The 413-ci wedge-head big-block was the corporation's largest V-8, powering such notable cars as the post-1958 letter-series Chrysler 300s.

Not surprisingly, the Max Wedge Mopars were sold without warranties. "The engine is designed for maximum acceleration from a standing start and should be excellently suited for special police pursuit work," explained Dodge Chief Engineer George W. Gibson in the 1962 press release announcing the 413 package.

One look at the spec sheet and Chrysler's trepidation at offering a warranty is understandable.

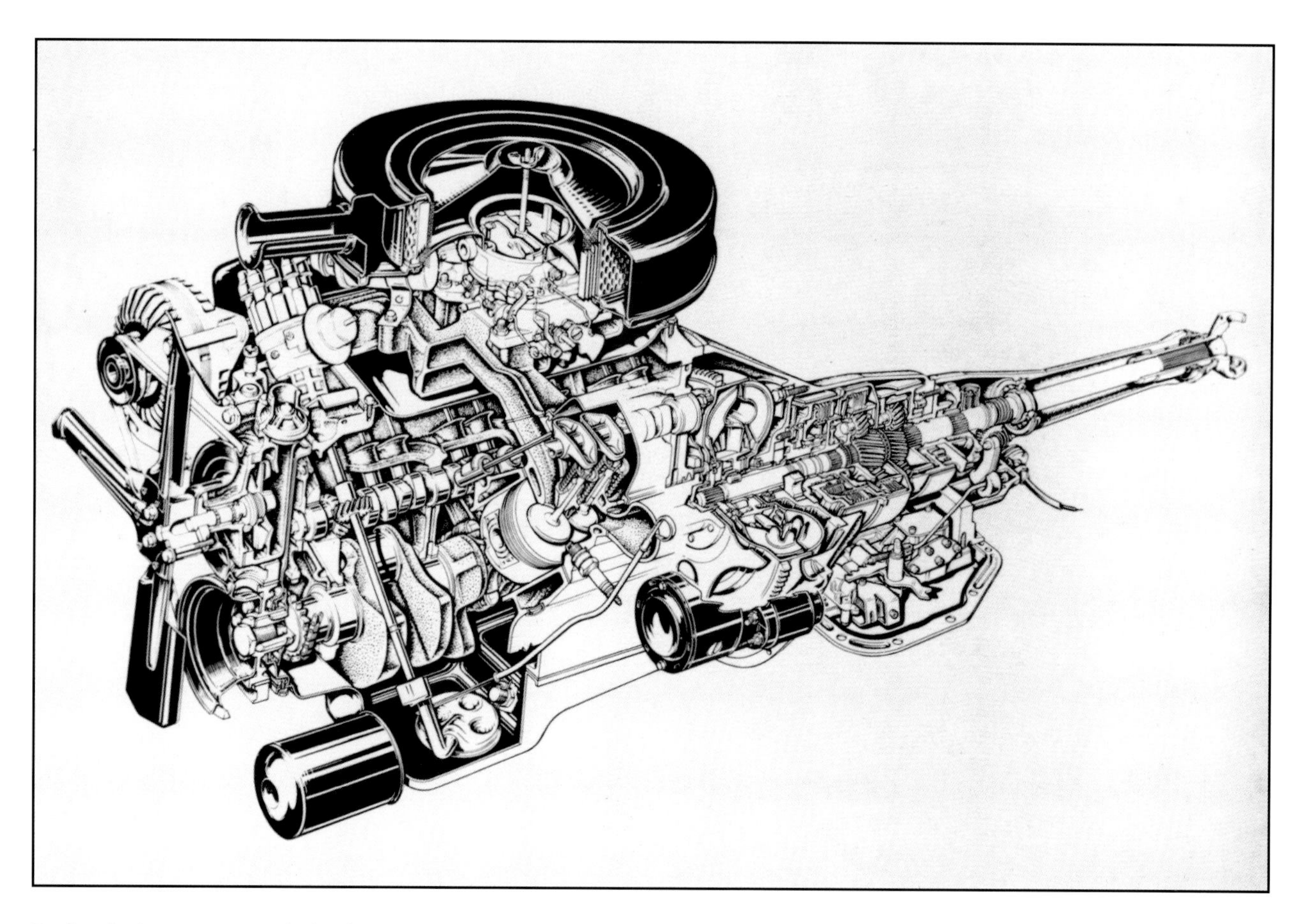

Dodge dealer Norm Kraus helped convince Dodge to build a 383-powered Dart by building his own prototype, and suddenly the Dart was in the big leagues. The 383-ci big-block, along with the 361 V-8 and 400 V-8, belonged to the "B" engine family. The "RB" big-block engine, like the 413, 426 Wedge, and 440, had a raised block to accommodate its longer stroke. The RB V-8 also had a larger main journal diameter. The B and RB big-blocks are referred to as "Wedge" engines because of their wedge-shaped combustion chambers.

As the Factory Experimental (F/X) classes evolved into the rock 'em, sock 'em Funny Car class, speeds got higher and so did the danger. The volatile nitro-methane Funny Cars were always at risk of bursting into flame, making the use of fire suits a necessity. Their cartoonish bodywork and incredible speeds made the Funny Cars popular for match racing, which often paid better than straight-up class competition. Here, Gene Snow lights 'em up in a 1968 Dart GTS. *Photo courtesy Drag Racing Memories*

The 413 Max Wedge utilized a one-piece aluminum intake manifold with short-ram, 15-inch runners, two four-barrel carburetors, high-capacity fuel pump, and high-flow air cleaners. The cylinder heads had a 25 percent larger port area than the standard 413, and the heat crossover passage was eliminated. The valves were sized large, at 2.08 inches on the intake side and 1.88 inches on the exhaust.

The label "heavy-duty" could be applied to just about every piece of the engine's guts, especially the forged steel crankshaft and connecting rods. With its forged aluminum pistons, the "base" Max Wedge 413 ran an 11.0:1 compression ratio. The engine was sold in two versions, the base 410-horsepower tune, and an even higher compression 420-horsepower edition. The rear suspensions of these cars were set up for maximum traction.

For 1963 Chrysler threw even more gasoline on the fire with the introduction of an even larger Max Wedge racing V-8, coming in at 426 cubic inches. The Max Wedge 426 was also sold in two states of tune, one at 415 horsepower and the ultra-high-compression version at 425 horsepower.

The Super Stock Hemi Mopars blasted out of the factories beginning in mid-1964. Although distantly related to both the 1950s-era Hemi engines and the wedge-head big-blocks, the 426 Hemi was definitely its own animal. The Hemi name was derived from the engine's hemispherical combustion chambers, a configuration that promoted good power production. The first 426 Hemis used cast-iron cylinder heads (although later versions used aluminum), a solid lifter cam, tubular steel headers, and ran a 12.5:1 compression ratio. The 1964 Super Stock Hemis were built with twin Holley four-barrel carburetors for drag racing or a single four-barrel for use in stock car racing.

Besides the race-tuned 426 Hemi engine, most of these early cars featured lightweight aluminum sheet metal and the extra heavy-duty A-833 four-speed transmission. When Chrysler built a factory race car, it put it all out on the track.

For 1965 Chrysler carried the factory Super Stock concept even further down the road of excess. The corporation actually produced a brief run of altered-wheelbase cars from the factory, destined for A/FX competition. These cars were built with the rear axle positioned several inches forward and the front wheels moved back, the better to achieve optimal weight transfer on launch. Besides the altered-wheelbase racers, Chrysler produced a run of ultralight Hemi Coronets and Belvedere I Super Stockers. Pounds were shed through the use of thin-gauge steel sheet metal and barely-there glass.

After declining to build a race-only Super Stock model for 1966, both Dodge and Plymouth returned to the track in 1967 with new Super Stock offerings. Both were more tame than the previous Super Stockers, although "tame" is a relative term here. The Dodge Coronet Super Stocker carried a WO23 identification code, the Plymouth Belvedere II an RO23 identification. These factory racers were built with the street Hemi, although the engines were bolstered with a modified intake manifold, transistorized ignition, and a functional thin-gauge steel hood scoop with sealed air cleaner assembly.

The 1967 editions, intended for A/Stock classification, weren't fitted with the aluminum sheet metal and Plexiglas of earlier super Stock Mopars, but did feature a few weight-saving modifications. They had the windshield wiper assembly removed, and the heater components were deleted. A monstrous trunk-mounted battery helped with weight distribution.

The Dart's turn, and the Barracuda's, arrived in early 1968. The potent combination of maximum engine and minimum weight proved irresistible to engineers and racers alike. The effort was spearheaded by Chrysler engineer Dick Maxwell, who had been responsible for many of the Super Stock Mopars, and engineers Bob Tarozzi and Larry Nolton. A 1967 Barracuda was used as the test mule.

A letter announcing the Hemi A-bodies was sent to dealers on February 20, 1968, with the cars released on March 4.

Although the cars were destined for a high-profile use, they represented a very small-scale project. To keep these unique race cars from cluttering up the assembly line, all Hemi Darts and Barracudas were built at Hurst Performance Research, a company increasingly well-equipped for the construction of limited-run performance cars. Dick Chrysler led the Hurst side of things.

Creating the Hemi Darts and Barracudas entailed a lot more than just tossing a Hemi between the fenders. Fiberglass front fenders and a fiberglass hood with scoop replaced the stock steel pieces, and the doors were acid dipped for further weight reduction. Additional pounds were saved through the use of Plexiglas windows with strap pull-ups, eliminating the window regulators. The rear seats were removed, and the stock front buckets were replaced with lightweight, low-back front buckets from the A-100 vans. Even the seat brackets were drilled to reduce weight. The brake master cylinder was relocated to allow room for the Hemi's massive valve covers. The battery was moved to the trunk for better weight distribution. Since the cars were destined for use on the drag strip, with all the custom painting and sponsor identification that entails, they were shipped unpainted from the factory. Only a primer coat covered the bare sheet metal.

The A-body edition of the 426 Hemi used dual 650-cfm Holley four-barrels on a cross ram manifold. The cars came standard with Hooker Headers and the exhaust pipes barely extended past the mufflers. Horsepower for the racing Hemi engine was still rated at a laughable 425 by the factory, the same as the street Hemi. A quick trip to the drag strip to witness Super Stock Darts running 10-second quarter-mile times, after only mild tuning, gave a more realistic sense of the Dart's true power output. Many racers consider the Hemi Dart and Barracuda to be the ultimate Super Stock Mopar, thanks to the cars' maximum power and minimum weight.

Best numbers indicate 80 1968 Super Stock Hemi Darts were built, and another 70 1968 Hemi Barracudas, although the totals are still fuzzy. Neither Hurst nor the lucky racers involved considered their Super Stock Dart efforts as the creation of a collectible classic. The Hemi-powered hot rods were merely this year's race car, not future auction bait.

Fittingly, a large number of the Darts were sold through Grand Spaulding Dodge in Chicago, the high-performance hotbed. "We told people that was our special!" recalled "Mr. Norm" Kraus, laughing.

Mopar PERFORMANCE
Koffel's Place
NHRA
S/ST
340
Minor's
Valvoline
GOODYEAR
EAGLE
SUNOCO
RACE FUELS
MSD
IGNITION
ROSS
racing pistons
JEG'S
HEDMAN
HUSLER
Mallory
PRO WIRE
B&M
SHIFTER
Mopar
PERFORMANCE PARTS
Valve Covers
Cylinder Heads
Mopar
PERFORMANCE PARTS
Ignition ECU
Engine Assembly
FEDERAL
MOGUL
B&G
SIMPSON
Holley
WELD
Auto Meter

The Dart and Valiant remain popular choices among the Mopar crowd, even 25 years after the cars have gone out of production. Due to the huge number produced, racers will be able to pick up inexpensive A-body cars for raw material for years to come.

With its relatively slick bodywork, the Duster was a good candidate for the Funny Car class. Tom "Mongoose" McEwen was one of the best-known Duster racers in the 1970s. *Photo courtesy Drag Racing Memories*

"By doing that, it sold the cars quicker. They wanted the 'specials' on the Grand Spaulding."

Grand Spaulding's reputation for setting up performance cars properly drew racers from around the country. "We did a complete car," Kraus said. "We did the carburetors, the hi-rises, the manifold, we did a competition valve job, we put in a gear, we put in an oiling system, we'd quick-shift the transmission, we put the traction bar on, we put the wheels, the tires on. We just went through everything. The cars went out. And all over the country the cars ran within one-thousandth of one another."

As the epicenter for Dodge high performance, the dealership did a booming business in Hemi Darts, until one night when the Grand Spaulding Hemi train was derailed.

"In August of '68 we had a fire in our dealership," Kraus recalled. "At that time, I don't know how many of them were delivered, but we had 18 of them in the stalls being set up. They would not take those cars unless we set them up."

Fortunately the fire was in a different part of the building from the waiting Hemi Darts. "We did finish those but we couldn't do any more," Kraus said.

"The facilities were really torn up. There was no roof on the front part of the dealership, and so forth."

The Super Stock Darts were generally purchased by serious racers, such as Bill Flynn, who made a racing name for himself with a series of "Yankee Peddler" Mopars, including one of the Hemi Darts. Californian Dick Landy was best known for his Charger race cars, but also competed in a Hemi Dart. (Prior to that, Landy had fielded an injected 1966 Dart in the fledgling Funny Car class. After 1966 he left the Funny Cars behind for Super Stock competition, and later Pro Stock.)

Landy is also known for the series of performance clinics he gave during the late 1960s and early 1970s. The performance clinic program, backed by the factory, was a traveling series of lectures and Q&A sessions held at Dodge dealers around the country. The clinics introduced a lot of would-be racers to the factory performance parts program, and helped raise the profile of the Direct Connection performance parts operation tremendously. Besides his racing and public relations feats, his Dick Landy Industries in Northridge, California, has been a steady supplier of race-ready 426 Hemi V-8s and restored Hemis as well.

Perhaps the most famous of the 1968 Super Stock Hemi A-bodies was the Barracuda of Ronnie Sox and Buddy Martin. They received the first Super Stock Hemi Barracuda, then another, and used them to good effect, racking up several victories. They also ran the Plymouth counterpart to Dick Landy's performance clinic, spreading the factory muscle gospel at Chrysler-Plymouth dealerships.

Of course, racers hit upon the idea of shoveling a Hemi in the A-body cars well before, and after, the factory contracted with Hurst to build the Super Stock specials. One such high-profile Hemi A-body was the 1965 Barracuda built and drag raced by stock car legend Richard Petty. When NASCAR disallowed the new 426 Hemi after the 1964 racing season, claiming it was not a "stock" production car engine, Chrysler picked up its football and went home—to drag racing. Whatever NASCAR thought, the Hemi had found a home in NHRA and AHRA drag racing. Richard Petty's crew built the "43 Junior" Barracuda (so named in a nod to Petty's traditional number used on the larger stock car Plymouths), and Petty campaigned it with some success.

In the 1970s

When Plymouth released the Duster in 1970, the new semifastback offered the same advantages (and disadvantages) as the Dart and A-body Barracuda—it was cheap, light, and available with a high-winding, powerful 340 V-8. It was also better looking than any Valiant in recent memory.

The Duster rode on the traditional Valiant 108-inch wheelbase, 3 inches shorter than the Dart's. Although not of much concern to the average racer, the Duster at least looked more aerodynamic than the Dart.

Some drag racers embraced the Duster as they had earlier Mopars. Among the prominent Duster racers were Arlen Vanke, Butch Leal, Tom "Mongoose" McEwen, and the Sox and Martin team. Most of these high-profile teams ran Dusters powered by the 426 Hemi in either the Funny Car or new Pro Stock classes.

As the 1970s progressed, however, corporate support for drag racing waned, especially as Chrysler's financial condition worsened later in the decade. The Dusters remained popular in the sportsman classes, but the days of the factory-supported drag racing heroes were over—at least for a while.

Some of the factory support shifted over to circle tracks. One of the last factory-sponsored A-body racing efforts revolved around the Dart Sport "Kit Car" Racing package. In the early to mid-1970s, Dodge engineered complete, ready-to-build Dart race car packages for the circle track racer.

"It's called a 'kit car' because anyone can purchase the packaged components and assemble them, just as youngsters put together scale models from a hobby shop," said Larry Rathgeb, manager of stock car programs for Chrysler, in a press release. "The buyer will be responsible for welding the parts and most will assemble their own engines. But, the beauty of the program is that we have taken the mystery out of building a stock car."

Engineered by Chrysler, the Dart kit pieces and cars were built at Richard Petty Engineering. Richard Petty even drove the Dart kit cars in a handful of races, although "The King" is better remembered for other chapters in his racing career.

The number of Dart kit cars built is unknown, but Chrysler felt the program was viable enough to continue running it through the late 1970s. Chrysler also produced Challenger kit cars, and later Dodge Aspen and Plymouth Volare kits.

The Dart, Demon, and Duster may have represented the cheap seats in the 1960s musclecar arena, but the A-bodies held up their end. The cars were showered with nearly as much performance engineering as their larger, more glamorous corporate stablemates, and the low price of entry allowed a lot of people to race who might not have been able to otherwise. That gift of opportunity might very well be the Dart's and Duster's enduring racing legacy.

383
FOUR BARREL

Appendices

Dodge Dart GTS/Swinger 340/Dart Sport 340 Production Numbers

Year/Model	Production Number
1967 GTS	457
1968 GTS	8,745
1968 Hurst/Super Stock Hemi Dart	80
1969 GTS	6,702
1969 Dart Swinger 340	16,637
1970 Dart Swinger 340	13,785
1971 Demon 340	10,098
1972 Demon 340	8,700
1973 Dart Sport 340	11,315
1974 Dart Sport 360	3,951

High-Performance Plymouth Duster Numbers

Year/Model	Production Number
1970 Duster 340	24,817
1971 Duster 340	12,886
1972 Duster 340	15,681
1973 Duster 340	15,731
1974 Duster 360	3,969
1975 Duster 360	1,421
1976 Duster 360	N/A

Detail of air cleaner, 1967 Dart GTS 383.

Base Prices, High-Performance Dodge Darts

Year/Model	Base Price
1964 Dart GT V-8 HT	$2,426
1964 Dart GT V-8 Convt.	2,644
1965 Dart GT V-8 HT	2,468
1965 Dart GT V-8 Convt.	2,687
1966 Dart GT V-8 HT Coupe	2,545
1966 Dart GT V-8 Convt.	2,828
1967 Dart GT V-8 HT Coupe	2,627
1967 Dart GT V-8 Convt.	2,860
1968 GTS HT Coupe	3,163
1968 GTS Convt.	3,383
1969 GTS HT Coupe	3,226
1969 GTS Convt.	3,419
1969 Swinger 340	2,836
1970 Swinger 340	2,808
1971 Demon 340	2,721
1972 Demon 340	2,759
1973 Dart Sport 340	2,853
1974 Dart Sport 340	3,320
1975 Dart Sport 360	4,014
1976 Dart Sport 360	*

* The Dart Sport 360 was dropped as a separate model after 1975. A $376 360-ci V-8, four-barrel, dual exhaust engine option could be ordered on the $3,370 Dart Sport V-8 with automatic transmission (except in California).

Base Prices (factory A.D.P.), High-Performance Plymouth Dusters

Year/Model	Base Price
1970 Duster 340	$2,547
1971 Duster 340	2,703
1972 Duster 340	2,742
1973 Duster 340	2,822
1974 Duster 360	3,288
1975 Duster 360	3,979
1976 Duster 360	*

* In 1976 Plymouth did not offer a separate "Duster 360" model, although the 360 four-barrel, dual exhaust V-8 was a $392 option available on the $3,353 Duster V-8 with automatic transmission (except in California).

Specifications, High-Performance Engines

273-ci High-Performance V-8 (available 1965–67)

Bore: ....3.63 in.
Stroke: ....3.31 in.
Compression Ratio: 10.5:1
Carburetion: ....single Carter AFB 4-barrel
..............................(single Holley 4-barrel, 1966 D/Dart)
Horsepower: ....235 at 5,200 rpm
....275 (1966 D/Dart)
Torque: ....280 at 4,000 rpm

340-ci V-8 (available 1968–73)

Bore: ....4.04 in.
Stroke: ....3.31 in.
Compression Ratio: 10.5:1 (1968–71)
....8.5:1 (1972–73)
Carburetion: ....single Carter AVS 4-barrel
Horsepower: ....275 at 5,000 rpm (1968–71)
.................................240 at 4,800 rpm (net, 1972-1973)
Torque: ....340 lb-ft at 3,200 rpm (1968-1971)
..............................290 lb-ft at 3,600 rpm (net, 1972–73)

383-ci V-8 (available 1967–69)

Bore: ....4.25 in.
Stroke: ....3.38 in.
Compression Ratio: 10.0:1
Carburetion: ....single Carter AVS 4-barrel
Horsepower: ....280 at 4,200 rpm (1967)
....300 at 4,400 rpm (1968)
....330 at 5,200 rpm (1969)
Torque: ....425 lb-ft at 5,200 rpm (1967)
.....................................400 lb-ft at 2,400 rpm (1968)
.....................................425 lb-ft at 3,400 rpm (1969)

426-ci Hemi V-8 (Super Stock Dart and Barracuda built by Hurst, available 1968)

Bore: ....4.25 in.
Stroke: ....3.75 in.
Compression Ratio: 12.5:1
Carburetion:............................dual 650-cfm Holley 4-barrel
Horsepower: ....425 at 5,000 rpm*
Torque: ....490 ft-lb at 4,000 rpm

* The "Super Stock" racing engines were considered underrated at 425 horsepower. Later street versions of the Hemi were rated at the same power level. Most estimates put the true figure for the race Hemi somewhere north of 500 horsepower.

Specifications, High-Performance Engines

440-ci V-8 (Hurst-built, 1969)

Bore: ....4.32 in.
Stroke: ....3.75 in.
Compression Ratio: 10.0:1
Carburetion: ....single Carter AVS 4-barrel
Horsepower: ....375 at 4,600 rpm
Torque: ....480 ft-lb at 3,200 rpm

360-ci V-8 (available 1974–76)

Bore: ....4.00 in.
Stroke: ....3.58 in.
Compression Ratio: 8.4:1 (1976)
Carburetion: ....single Carter 4-barrel
Horsepower: ....245 at 4,800 rpm (net, 1974)
..................................230 at 4,000 rpm (net, 1975)
..................................220 at 4,400 rpm (net, 1976)
Torque: ....320 lb-ft at 3,600 rpm (net, 1974)
...............................300 lb-ft at 3,200 rpm (net, 1975)
...............................280 lb-ft at 3,200 rpm (net, 1976)

Index